Lecture Notes in Mechanical Engineering

ABCM Series on Mechanical Sciences and Engineering

This series publishes selected papers as well as full proceedings of events organized and/or promoted by the Brazilian Society for Mechanical Sciences and Engineering (ABCM) on an international level. These include the International Congress of Mechanical Engineering (COBEM) and the International Symposium on Dynamic Problems of Mechanics (DINAME), among others.

Milla Caroline Gomes · Déborah de Oliveira ·
Maksym Ziberov
Editors

Innovations in Brazilian Machining

Highlights from the XXVI Machining Colloquium

ABCM Associação Brasileira de Engenharia e Ciências Mecânicas

Springer

Editors
Milla Caroline Gomes
Department of Manufacturing
and Materials Engineering
University of Campinas-UNICAMP
Campinas, Brazil

Déborah de Oliveira
Department of Mechanical Engineering
University of Brasília—UnB
Brasília, Brazil

Maksym Ziberov
Department of Mechanical Engineering
University of Brasília—UnB
Brasília, Brazil

ISSN 2195-4356 ISSN 2195-4364 (electronic)
Lecture Notes in Mechanical Engineering
ISSN 2662-3021 ISSN 2662-303X (electronic)
ABCM Series on Mechanical Sciences and Engineering
ISBN 978-3-031-92818-5 ISBN 978-3-031-92819-2 (eBook)
https://doi.org/10.1007/978-3-031-92819-2

This Springer imprint is published by the registered company Springer Nature Switzerland AG
The registered company address is: Gewerbestrasse 11, 6330 Cham, Switzerland

Preface

The Colloquium on Machining is a Brazilian academic conference that began as a local event in the city of Uberlândia-MG, at the Federal University of Uberlândia (UFU), with its first edition held in 1993. Its primary goal was to promote the dissemination and discussion of ongoing research in machining. Due to the increased credibility it gained, and the space created for dialogue on the advances and challenges in the area, the event grew, attracting participation from both the academic and industrial communities. Over time, it became an annual event with national reach. In addition to Uberlândia-MG, the Colloquium has been held in several other cities, including Belo Horizonte-MG, Goiânia-GO, Juiz de Fora-MG, Viçosa-MG, Natal-RN, Ouro Preto-MG, Uberaba-MG, Curitiba-PR, and Brusque-SC. In 2024, the 26th edition was held for the first time in the state of São Paulo, in Campinas, at the State University of Campinas (UNICAMP), from November 25 to 27.

The XXVI Machining Colloquium was organized by professors from different institutions, with the following participants: Dr. Milla Caroline Gomes (UNICAMP) as Chair, Dr. Álisson Rocha Machado (PUCPR), Dr. Amauri Hassui (UNICAMP), Dr. Daniel Iwao Suyama (UNICAMP), Dr. Marcio Bacci da Silva (UFU) and Dr. Rodrigo Blödorn (UNIFEBE) as links and actions with the industry, Dr. Maksym Ziberov (UnB) and Dr. Déborah de Oliveira (UnB) responsible for the Scientific Committee, Dr. Raphael Lima de Paiva (UFPI) and Dr. Aline Gonçalves dos Santo (UFCAT) as Treasurers, Dr. Suzana Regina Moro (UNICAMP) and Dr. Paula Fernanda da Silva Farina (UNICAMP) as Secretaries. In addition to the professors, UNICAMP undergraduate students were involved: Beatriz Dos Santos Bragaglia, Thiago Elias Gomes, Pedro Araújo Fernandes and Laysa Ramos Orlovas.

During the three days of the 26th Machining Colloquium, 120 participants from five different states and the Federal Districts took part in the event's program. The schedule included 7 lectures, 1 roundtable discussion, 2 workshops, and the presentation of 56 technical papers, spread across 9 distinct areas of knowledge: Automation and Control Technology of Machining Processes, Process Parameter Improvement, Metrology in Machining Processes, Micromachining, Machine Tool Design, Manufacturing Systems and Planning, Surface Texture and Integrity, Material Machinability, and Green Machining.

Of the 56 papers approved and presented at the Colloquium, 21 were carefully selected to be included in this book. Among the selected works are those that were awarded as the best in their respective fields. It is important to note that the papers were chosen to reflect the key topics discussed during the event. Therefore, the book is organized as follows: Chapter "A Comprehensive Overview on the Machining Colloquium Demography" presents a demographic study to provide a better understanding of the impact of the Machining Colloquium. Chapters "Analysis of Polyamide Roughness After Grinding with Different Workspeeds and Radial Depths of Cut"–"Investigation of the Impact of Different Trajectory Angles on the Milling of ER70S-6 Parts Obtained by WAAM-CMT" focus on Process Parameter Improvement. Chapters "Calculation of Measurement Uncertainty Associated with Ra Roughness of FC 250 Gray Cast Iron After Grinding"–"Measurement Uncertainty Calculation Associated With Ra Roughness of Titanium Alloy After Grinding" address Metrology in Machining Processes. Next, Chapters "Analysis of Surface Roughness in Micromilling AISI 316L Stainless Steel with Variation of Feed per Tooth"–"Tool Wear in Micro End Milling of Ultrafine Grain Steels" are dedicated to Micromachining. Chapters "Application of Heat Map as Imaging Treatment to Enhance Surface Characteristics Visualization and Analysis for Milled Wire and Arc Additive Manufacturing Parts" and "Machining of Hybrid Rocket Engine Nozzle Manufactured by Wire and Arc Additive Manufacturing" cover, respectively, Machine Tool Design and Manufacturing Systems and Planning. Chapters "Effect of the Wire Electrical Discharge Machining and Electropolishing on Surface Quality of the Nd-Fe-B Magnet"–"Texture Behavior in Ni-Cr-Fe Alloys by Different Lateral Infeed Machining in WEDM Process" discuss Surface Texture and Integrity. Finally, Chapter "Energy Efficiency in Machine Tools: Impact of Cutting Parameters on Sinusoidal Paths" deals with Green Machining.

The book covers current topics and research conducted by specialists in the field of Machining in Brazil. Therefore, it is recommended for researchers, engineers, and undergraduate and graduate students who wish to stay up to date on the key issues in the field.

Campinas, Brazil — Milla Caroline Gomes
Brasília, Brazil — Déborah de Oliveira
Brasília, Brazil — Maksym Ziberov

Contents

A Comprehensive Overview on the Machining Colloquium Demography

Milla Caroline Gomes, Maksym Ziberov, Aline Gonçalves dos Santos, Raphael Lima de Paiva, and Déborah de Oliveira

Abstract The Machining Colloquium is a Brazilian conference on machining, established in 1993 by the Laboratory for Teaching and Research in Machining (LEPU) at the Federal University of Uberlândia (UFU). Its primary goal is to bring together researchers to discuss both completed and ongoing projects, fostering networking and partnerships. While the conference attracts participants from various states, a significant presence of students and professors from Minas Gerais has been noted, likely due to the conference's origins and its occasional hosting in the state. A demographic study was conducted through a questionnaire to explore these patterns. The analysis revealed that most participants are aged 31–40, suggesting a majority of graduate students or early-career researchers and professors. Males comprised 84% of attendees, and the ethnic distribution mirrored national undergraduate trends: 66% white, 28% pardo, 4% black, and 2% yellow. Finally, the state of Minas Gerais stood out for being responsible for the formation of more than 50% of the attendees, regardless of the degree. This was associated with both the geographical location and the impact of LEPU on the formation of human resources in machining research.

Keywords Machining · Conference · Education · Demography

M. C. Gomes (✉)
Department of Manufacturing Engineering and Materials, University of Campinas, Campinas, SP, Brazil
e-mail: millagm@unicamp.br

M. Ziberov · D. de Oliveira
Department of Mechanical Engineering, University of Brasilia, Brasilia, DF, Brazil

A. G. dos Santos
Production Engineering Department, Federal University of Catalão, Catalao, GO, Brazil

R. L. de Paiva
School of Mechanical Engineering, Federal University of Piauí, Teresina, PI, Brazil

M. C. Gomes et al. (eds.), *Innovations in Brazilian Machining*, Lecture Notes in Mechanical Engineering, https://doi.org/10.1007/978-3-031-92819-2_1

1 Introduction

Many engineering industries rely on machining operations to manufacture their products. Machining processes offer several advantages, including the ability to machine various materials, create complex shapes and special geometric features; high dimensional accuracy; and excellent surface finish [1]. The importance of the machining process for industry has led to an increasing depth of study and discussion over the years. In this context, conferences that aim to foster scientific production and promote meaningful debates on the subject stand out.

In this scenario, the Machining Colloquium was conceived, a technical-scientific event organized since 1993 by the Laboratory for Teaching and Research in Machining, in Portuguese Laboratório de Ensino e Pesquisa em Usinagem (LEPU), characterized as a local meeting of a specific nature, whose objective was to disseminate research developed in "Machining Processes" to the academic-scientific community of the Federal University of Uberlandia. Therefore, in the first edition the main audience were the students from LEPU, from undergraduate and graduate programs.

The Machining Colloquium has grown significantly, attracting researchers from various institutions. This expansion led to its first edition outside the Federal University of Uberlândia during the 8th event in Belo Horizonte. Today, it is recognized as the largest machining conference in Brazil, with recent editions featuring around 60 works about machining, an interesting number when comparing to the Brazilian Manufacturing Engineering Congress (COBEF), a national event that in the last edition, 2023, counted with 406 attendees, from 17 different states and the Federal District of Brazil and 8 different countries, totaling 227 technical works. Besides being a success and a traditional conference in manufacturing, the relevance of the Machining Colloquium stands out when it is noticed that COBEF has 10 different areas of knowledge: Additive Manufacturing, Project and Management, Product Engineering, Processes with Material Removal, Computational Vision, Control Systems, Robotics, Sensors and Actuators, Materials Characterization and Processing, Metrology, Tribology, Union and Assembly Processes and Metallurgical Processes [2].

According to Dennis et al. [3] the success of a conference is associated with the ability of the organizing committee to overcome the difficulties in planning, funding and disclose. The authors mention that some of the difficulties remain in the fact that it is not a skill usually obtained in the undergraduate, however those events are considered important by the authors once they create the opportunity for information sharing and learning. Considering the unique characteristics of the Machining Colloquium, this work aims to comprehend which details can differ it from other conferences and evaluate the demographic impact of the conference, understanding its impact in the different Brazilian regions.

2 Materials and Methods

2.1 Previous Editions Data

To obtain the data of the previous conferences the official documents were collected, considering official sources such as: Lattes curriculum (official governmental curriculum for Brazilian researchers), Documents submitted to funding agencies, Institutional documents from the universities and Website from previous editions.

The data was collected direct from the sources mentioned and provided by the previous organizing committees from the respective edition. This step allowed to comprehend the evolution of the event regarding its evolution from local event, transitioning to regional until became a national conference in the light of the geographical distribution of the editions.

2.2 Participants Demography Data

Aiming to acquire the characteristics of the participants a questionary was applied, considering the terminologies of the Brazilian Institute of Geography and Statistics. Participants from editions prior to 2024 were invited to answer questions considering their origin, features and formation. The interface selected to apply the questionary was the Forms, from Microsoft, following the details on Table 1.

The selection of some of the questions was based on the work of Ascenso et al. [4] and Atikuzzaman and Ahmed [5], resulting in the following questions, the text in parentheses represents the possible answers for each question: 1. Name (Text—Name of the participant), 2. Age (number), 3. Gender (multiple choice: male, female, other), 4. Ethnicity (multiple choice: White, Black, Pardo, Yellow), 5. State of origin (Text—Name of the Brazilian states), 6. Did you graduate? (Yes, No, I'm currently an undergraduate), 7. In which institution did you graduate? (Text—Name of the institution), 8. Do you have a master's degree? (Yes, No, I'm currently pursuing my Master's), 9. In which institution did you obtain your master's? (Text—Name of the institution), 10. Do you have a PhD? (Yes, No, I'm currently pursuing my PhD), 11. In which institution did you obtain your PhD? (Text—Name of the institution), 12. Do you have a Post Doc? (Yes, No, I'm currently pursuing my Post Doc), 13. In which institution did you obtain your Post Doc? (Text—Name of the institution), and 14. Did you participate in the Machining Colloquium? (Yes, No).

On the data analysis item 3.1 (Table 1), were considered valid only the answers that were complete and without any random answers. After the screening the groups were evaluated in terms of the percentage of the characteristics and finally organized in a graphical representation that will be presented at item 3. In order to clarify for readers from different countries, below is a list of the main abbreviations:

Table 1 Methodology for data acquisition through questionnaire

Item	Method	Description
1. Questionnaire disclosure	1.1. Colloquium official website	Public access to the Forms link
	1.2. Colloquium mail-list	
	1.3. Machining research groups	Access to the Forms link focusing on the machining community
	1.4. Authors LinkedIn	
	1.5. Personally, to machining professors	Personal contact with the researchers, professors and students that participated previously
	1.6. From advisors to advisees	
2. Data collection	2.1. Forms responses	After 2 months of collecting the data was acquired for treatment
3. Data analysis	3.1. Data treating on Excel	Selection and classification of the answers
	3.2. Graphical representation	Elaboration of the graphical representation
	3.3. Discussion	Evaluation of the graphical results

- **Brazilian State**: ES—Espírito Santo, GO—Goiás, MG—Minas Gerais, PR—Paraná, RJ—Rio de Janeiro, RN—Rio Grande do Norte, SC—Santa Catarina, SP—São Paulo.
- **Federal University**: UNIFEI—Federal University of Itajubá, UFMG—Federal University of Minas Gerais, UFPE—Federal University of Pernambuco, UFRGS—Federal University of Rio Grande do Sul, UFSC—Federal University of Santa Catarina, UFSJ—Federal University of São João Del Rei, UFTM—Federal University of Triângulo Mineiro, UFU—Federal University of Uberlândia, UnB—University of Brasília, ITA—Institute of Aeronautical Technology, UTFPR—Technological Federal University of Paraná.
- **Federal Institute**: CEFET-MG—Federal Center for Technological Education of Minas Gerais, CEFET-GO—Federal Center for Technological Education of Goiás, IFES—Federal Institute of Espírito Santo.
- **Estate University**: UNESP—State University Júlio de Mesquita Filho, UNICAMP—Campinas State University, USP—University of São Paulo.
- **Catholic University**: PUCMG—Catholic University of Minas Gerais, PUCPR—Catholic University of Paraná.
- **Institute of the Brazilian Federal Administration**: IBGE—Brazilian Institute of Geography and Statistics, INEP—National Institute of Educational Studies and Research Anísio Teixeira.
- **Conference of the Brazilian Association of Engineering and Mechanical Sciences**: COBEF—Brazilian Manufacturing Engineering Congress.

3 Results and Discussion

3.1 Contextualizing the Machining Colloquium by the Previous Editions

The Machining Colloquium has been held annually since 1993 at various institutions across Brazil. The event began at the UFU in Uberlândia, Minas Gerais (MG), in the Southeast region, where it took place for the first seven editions. In 2002 the first edition outside the UFU was in Belo Horizonte, the capital of the state, indicating an expansion on the public but the presence of a great regionalism of the impact. The 8th Machining Colloquium was held in Belo Horizonte-MG, at PUCMG. Following this, it returned to the Federal University of Uberlândia for the 9th edition and was sponsored by the Instituto Fabrica do Milenio (IFM). In 2004 it was hosted by CEFET-MG in Belo Horizonte. The positive results of the event had repercussions at other research institutions in the region. For this reason, the 11th Machining Colloquium was held by the UFU together with the CEFET-GO, Midwest region. The event was attended by other institutions, such as the Technological Faculty of SENAI.

On the following years the event starts to take place on Uberlandia and other cities from MG, such as Juiz de Fora (2011) and Viçosa (2012). However, on 2015 the Colloquium occurred in Natal, Rio Grande do Norte (RN) state, located in the northeast of the country, leading to a big expansion on the public. An important information to highlight is that the committee for the Natal edition was formed by professors who were students at LEPU during the graduate program and had previously participated in the event previously. According to the data, this characteristic was also present on the committees of the following editions that took place on Ouro Preto—MG (2017) and Uberaba—MG (2018), with the conference chairs from 2016 to 2019 being advisees from the same supervisor from LEPU.

As other conferences the Machining Colloquium was affected by the COVID-19, and it is possible to notice a void in the periodicity of the event from 2019 to 2022. This occurred due to the decision to not make an online event, considering the main characteristic of the Colloquium being the possibility of interaction and discussion of the works. In 2022 with the possibility to return to face-to-face conferences it took place in Curitiba, Parana (PR) state, located on the south region. This new expansion on the public and the occurrence in a new region was attributed to the presence in the Catholic University of Parana of senior full professor that was one of the founders of the Colloquium and the LEPU itself. From there, the Colloquium could expand further south, to Brusque, in Santa Catarina (SC) state and finally being held in Campinas, São Paulo (SP) state, on the present edition.

Considering the continental territory of Brazil, the expansion of the local event through the years is an important factor to consider. To help to visualize, Fig. 1 was elaborate with the approximate location of the cities that held the event. On Fig. 1, one can note that besides the expansion there are still a concentrated region where it is more probable that the event will take place.

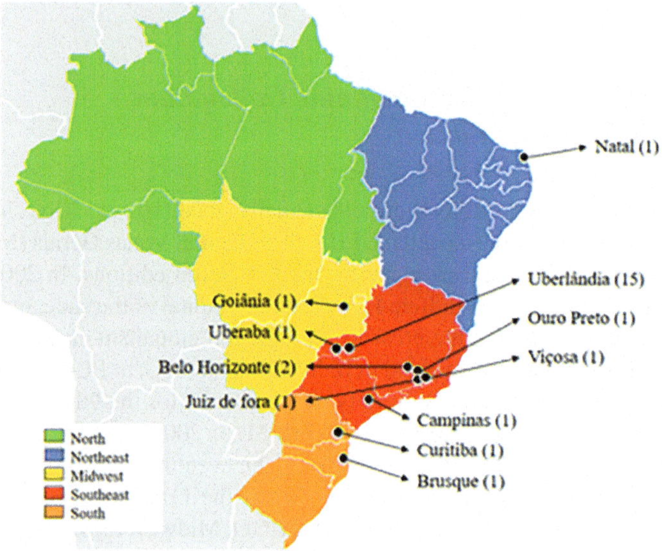

Fig. 1 Geographical distribution of the Machining Colloquium across Brazilian regions, cities and frequencies

When analyzing data from questionnaires, projects, marketing, and websites of various editions, some defining characteristics of the Colloquium consistently emerge, regardless of the organizing committee's style:

- Audience: Professors, graduate (master's and doctoral) and undergraduate students, engineers, and technicians focused on "Machining Processes."
- Objective: To provide a forum for discussing machining research, enhancing participants' technical and scientific knowledge through lectures, mini-courses, round tables, poster sessions, and oral presentations.
- Submissions: Two categories—full-length articles and extended resumes—encourage student participation and discussion of early-stage work.
- Registration: Free for all participants.
- Organizing Committee: Predominantly graduate students, often including the Colloquium chair, with undergraduates supporting event tasks. Senior professors act as supervisors/advisors, while early-career professors frequently serve as chairs or committee members. Committees typically include members with prior experience or ties to LEPU.

3.2 Participants Demography Data

In the light of the possible high impact of LEPU and the Colloquium in the region, understanding participant characteristics such as birthplace and educational background is essential to assess the conference's influence. In this sense, this section presents attendee data, from general demographics to educational profiles. Regarding age, Fig. 2 shows that most participants fall within the 31–35 range, followed by 35–40, suggesting they are primarily graduate students or early-career professors. The youngest respondent was 20 years old, the oldest 67, resulting in a 47-year age range.

Participants were asked to identify their gender and ethnicity based on categories from the Brazilian Institute of Geography and Statistics (IBGE), as shown in Fig. 3. The gender gap (Fig. 3a), with fewer women than men, reflects a trend in Engineering courses. According to the National Institute of Educational Studies and Research Anísio Teixeira (INEP), women represent about 36% of students in higher education courses related to engineering, production, and construction [6]. No participants identified itself as “other” genders. Most participants identified as white, followed by pardo (mixed origin, as defined by IBGE). Figure 3b shows a distribution consistent with the 2022 Higher Education Census, where graduates, regardless of the type of institution and the courses, were 45.8% white, 31.2% pardo, 7.1% black, 2.3% yellow or native, and 13.7% undeclared [7]. This suggests the conference is representative of the higher education population but not the general Brazilian population, which is 45.3% pardo, 43.5% white, 10.2% black, and 1.0% yellow or native [8].

Figure 4 presents the birthplace of attendees, showing that 65% are from the Southeast region, which includes São Paulo (SP), Minas Gerais (MG), Rio de Janeiro (RJ), and Espírito Santo (ES). All four states were represented, with MG having a

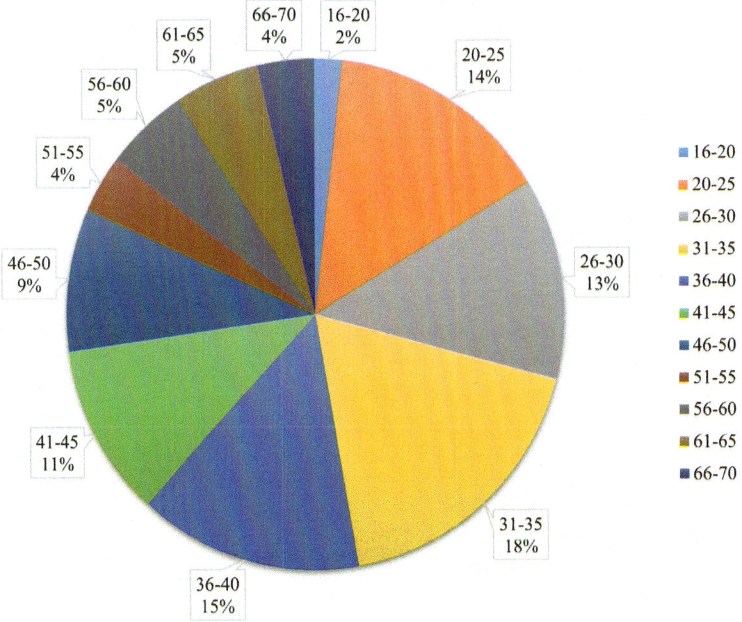

Fig. 2 Age ranges of Machining Colloquium participants

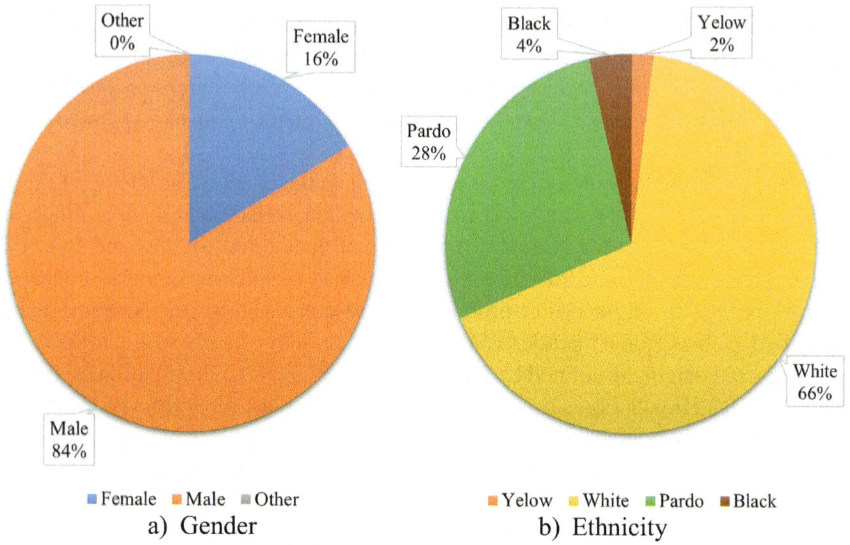

Fig. 3 Participants characteristics: **a** Gender and **b** Ethnicity

particularly strong presence. According to IBGE [8], SP has a population of over 44 million, MG 20 million, and RJ 16 million. However, this data should be viewed with caution, as many students relocate to study at specific institutions, especially Federal Universities, which provide free access to higher education.

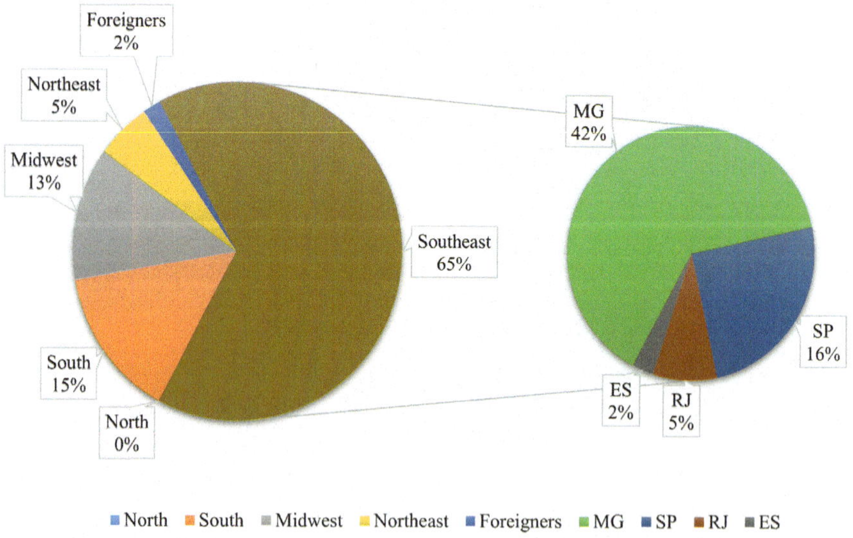

Fig. 4 Participants origin

Since some regions serve as higher education hubs, many participants likely graduated in cities different from their birthplace. Understanding where they graduated is crucial, as it impacts their interest in the Colloquium. Figure 5a confirms this: 43% of participants graduated from the UFU, where the event originated. The UnB had the second-highest representation, mainly from participants aged 16–20 and 21–25, most of whom are undergraduates. This is linked to two manufacturing professors at UnB, who are UFU alumni and former Colloquium participants and organizers. This indicates the event fosters a culture of academic and research engagement, encouraging student participation. Considering postgraduate students, the behavior is similar (Fig. 5b), UFU stood out followed by UNICAMP. This may be because UNICAMP is one of the important centers in research in machining in Brazil, when considering the universities. UFTM and PUCMG appear with 4%. The PUCMG is in city of Belo Horizonte, where the event occurred twice, and again the UnB. It can also be noted that similar to undergraduate students, there are a wide range of different institutions.

For the PhD and the Post Doc, Fig. 6, as expected, the range of universities reduced. For undergraduates there were 15 different universities, 14 for master's degree, 12 for the PhD and 10 for Post Doc. This was expected not only due to the availability of the courses but also due to the formation of the participants that not always possess PhD or higher education.

When the data is grouped by the states, as presented in Fig. 7, it is possible to observe the prevalence of participants which the formation took place in MG, for all the different stages of formation more than 50% of the attendees studied in Minas Gerais. Focusing on the undergraduate, Fig. 7a as mentioned, the presence of students from UnB is recent, started in the 2022 edition, on the other hand SC and SP are considered traditional hubs for engineering research. This can be confirmed by the master's data, Fig. 7b, with SP and SC being responsible for 19% and 15% of the attendees respectively. Finally for the PhD, the distribution becomes less wide and there is an increase on foreign institutions.

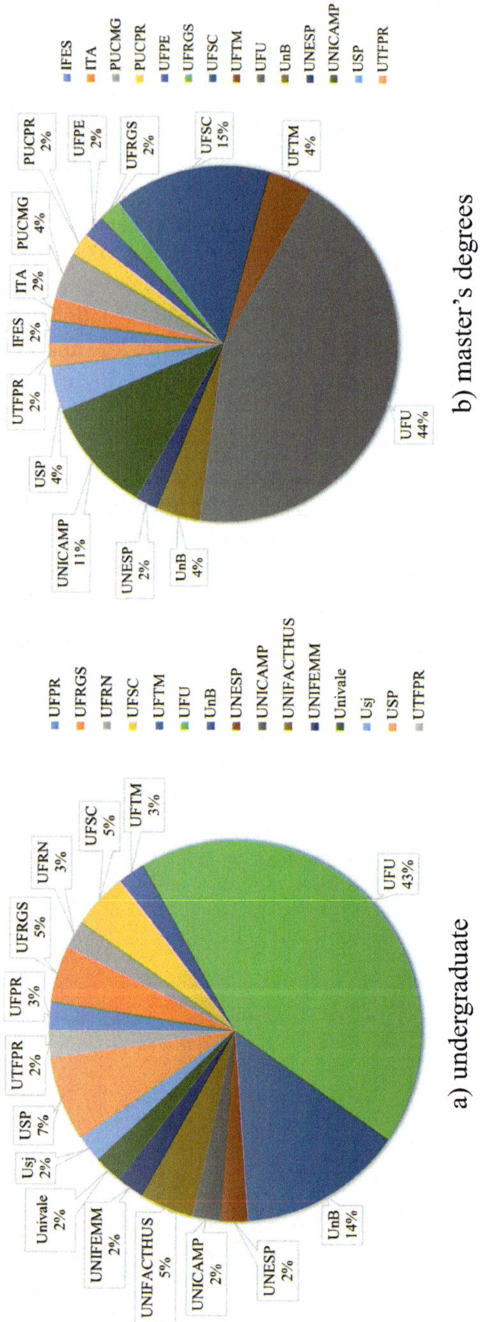

Fig. 5 Universities where participants completed their undergraduate and master's degrees

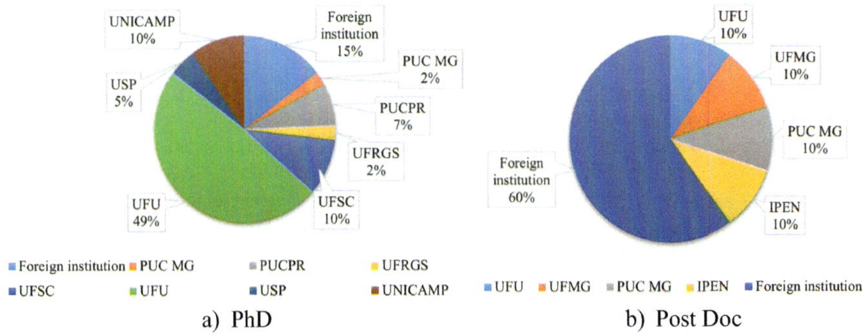

Fig. 6 Participants university of post graduate: **a** PhD and **b** Post Doc

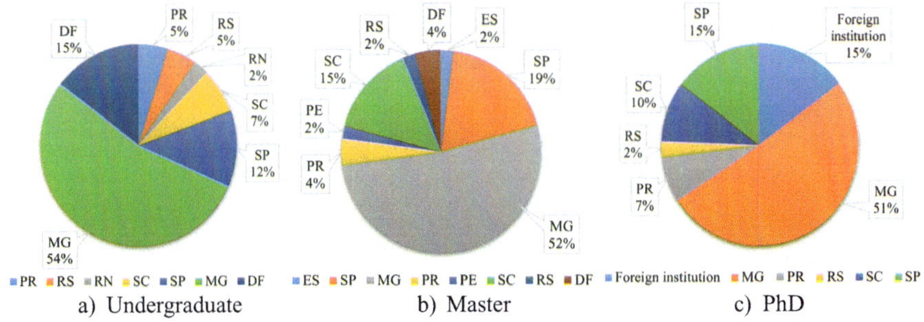

Fig. 7 Participants state of formation: **a** Undergraduate, **b** Master and **c** PhD

4 Conclusions

After the data collection on the participants of the Machining Colloquium and the analyses, the following conclusions could be drawn:

- The Machining Colloquium is a well-established conference in Brazil, held annually for over 30 years. It began as a local symposium, expanded to a regional conference at the 8th edition, and became a national event by the 19th edition.
- The majority of participants (31–40 years old) are likely graduate students, researchers, or early-career professors.
- Males make up 84% of participants, while females account for 16%, with no representation from other genders.
- In terms of ethnicity, 66% identified as white, followed by 28% as pardo, 4% as black, and 2% as yellow.
- Minas Gerais stands out, contributing over 50% of attendees, due to both its geographical proximity and LEPU's significant influence on machining research and human resource development.

References

1. Huda Z (2020) Machining processes and machines: fundamentals, analysis, and calculations, 1st edn. CRC Press. https://doi.org/10.1201/9781003081203
2. de Oliveira D, Ziberov M, Machado AR (2024) ABCM series on mechanical sciences and engineering. Springer Nature Switzerland. https://doi.org/10.1007/978-3-031-43555-3
3. Dennis S, Zuccaro J, McAllister C (2023) Conference program planning: Improving the communication approaches and evaluation of hybrid conference proposals. J Acad Librariansh 49(4):102721. https://doi.org/10.1016/j.acalib.2023.102721
4. Ascenso RMT, Luis L, Dias SS, Gonçalves D (2024) First glance into academics health literacy—socio-demographics. Procedia Comput Sci 239:1070–1077. https://doi.org/10.1016/j.procs.2024.06.271
5. Atikuzzaman MD, Ahmed ZSM (2023) Investigating the impact of demographic and academic variables on assessing students' perceived information literacy self-efficacy. J Acad Librariansh 49(4):102733. https://doi.org/10.1016/j.acalib.2023.102733
6. INEP (2024) Higher education census of 2022. https://download.inep.gov.br/educacao_superior/censo_superior/documentos/2021/apresentacao_censo_da_educacao_superior_2021.pdf
7. INEP (2024) Technical summary of the higher education census of 2022. https://download.inep.gov.br/publicacoes/institucionais/estatisticas_e_indicadores/resumo_tecnico_censo_educacao_superior_2022.pdf
8. IBGE (2023) Brazilian census of 2022—demographic census. https://censo2022.ibge.gov.br/panorama/indicadores.html

Analysis of Polyamide Roughness After Grinding with Different Workspeeds and Radial Depths of Cut

Jamille Kessia Araujo Martins, Marcus Vinícius de Oliveira, Josenilton dos Santos Lopes, Bruno Souza Abrão, Mark James Jackson, and Rosemar Batista da Silva

Abstract The automotive sector is increasingly adopting sustainable practices, emphasizing weight reduction technologies and electrification. This shift necessitates changes in materials and component design, highlighting polymers and fiber-reinforced materials for their emission reduction and recyclability. Polyamide 6, also known as Tecnil, has emerged as a key material due to its lightness, recyclability, and suitability for innovative methods like rapid prototyping and machining for small batches. Grinding is often the preferred method for achieving specific geometries, tolerances, and surface finishes, but there is limited literature on the cutting parameters for grinding polymers, especially polyamides, with conventional abrasive wheels. Due to their lower melting point, polymers require careful selection of cutting parameters, particularly radial depth of cut and workspeed, to manage heat generation and avoid defects like poor finish and thermal damage. This study evaluates the influence of radial depth of cut (15 and 45 μm) and workspeed (2.7 and 7.5 m/min) on the surface roughness of Polyamide 6 using a silicon carbide grinding wheel. Results indicate that all roughness parameters (Ra, Rz, Rq, and Rt) increase with higher radial depths of cut and workspeeds. Increasing the depth of cut from 15 to 45 μm raised Ra by up to 361.29% at 2.7 m/min and 353.54% at 7.5 m/min. Raising workspeed from 2.7 to 7.5 m/min increased Ra by 31.05% for 15 μm and 28.86% for 45 μm. Despite these increases, Ra remained below 1.6 μm, suitable for semi-finished automotive components. Optimal grinding conditions were achieved with a 15 μm depth of cut and 7.5 m/min workspeed, resulting in a minimum Ra of 0.325 μm. These findings guide parameter selection for balancing surface quality and grinding efficiency in Polyamide 6 manufacturing.

Keywords Grinding · Polyamide 6 · Roughness · Radial depth of cut · Workspeed

J. K. A. Martins (✉) · M. V. de Oliveira · J. dos Santos Lopes · B. S. Abrão · R. B. da Silva
Universidade Federal de Uberlândia, Uberlândia, MG, Brazil
e-mail: jamille.martins@ufu.br

M. J. Jackson
Kansas State University, Aerospace and Technology Campus, Salina, KS, USA

M. C. Gomes et al. (eds.), *Innovations in Brazilian Machining*, Lecture Notes in Mechanical Engineering, https://doi.org/10.1007/978-3-031-92819-2_2

1 Introduction

The quality of the final surface, defined in terms of roughness, defects, texture and waviness, is one of the main concerns in manufacturing as it directly influences production costs and the integrity of the final part [1]. In this context, grinding stands out as one of the most widely used machining processes to ensure the combination between low surface roughness values Ra in range of 0.1–1.6 μm with the International Tolerance (IT) from IT 06 IT 03. It is an abrasive machining process where material is removed by high cutting speeds ($V_s \geq 30$ m/s) [2, 3]. Measurement of surface roughness, especially the Ra and Rz parameters, allows one to evaluate the quality of a machined workpiece and is widely used to assess the performance of grinding processes [4]. The variation of cutting parameters, such as cutting speeds (V_s), workspeeds (V_w), and radial depths of cut (a_e) plays a significant role in the outcome of the ground surface affecting both the roughness and the surface integrity of the part [5]. Thus, measuring the roughness of materials like Polyamide 6 (also known as Tecnil) as a function of cutting parameters are a form of efficiency for evaluating the grinding process. The thermoplastic is widely used in industries that demand materials with high wear and abrasion resistance, especially under severe operating conditions, such as high speed grinding with different radial depths of cut [6]. The mechanical properties of Polyamide 6 can vary significantly with cutting conditions, particularly with grinding temperature, impacting the final roughness and texture of the part. A surface with low roughness is crucial in industrial applications that require precision and quality, such as in tribological and structural components where surface integrity directly influences the performance and durability of the final product. Rigorous control of surface roughness is essential to ensure that Polyamide 6 maintains its durability, dimensional stability, and suitability for critical applications, such as gears, bearings, and other parts that require high precision, for instance, parts that will be assembled into others [7]. Therefore, understanding these relationships is essential to optimizing the grinding process, ensuring the production of components with high precision and superior surface finish, and meeting modern industrial demands [8]. Thus, the study aims to investigate the influence of radial depths of cut (a_e) and workspeeds (V_w) on the roughness parameters Ra, Rz, Rq, and Rt during the surface grinding of Polyamide 6 with a silicon carbide grinding wheel (SiC).

2 Methodology

The experimental tests were conducted on a semi-automatic peripheral surface grinding machine grinder, model P36, manufactured by Mello, with 3 HP motor and 2400 rpm speed, a power of 2.25 kW, and a z-axis (vertical) resolution of the grinding wheel axis movement equal to 5 μm. A silicon carbide grinding wheel (SiC) with a 39C46KVK structure was used. It has the following dimensions: 239.0 mm (outer diameter) × 25.4 mm (width) × 255.0 mm (inner diameter). The material

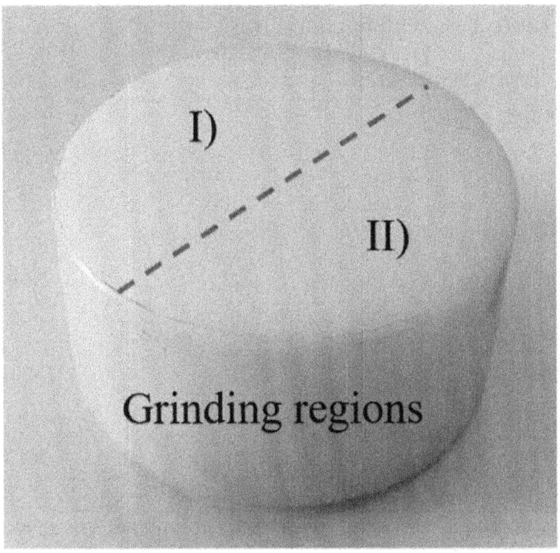

Fig. 1 Ground Polyamide 6 sample

ground was Polyamide 6 supplied in a circular disc shape with 50.6 mm (diameter) × 24.4 mm (height). A photo of the ground Polyamide 6 sample is shown in Fig. 1.

A vegetable-based semi-synthetic VASCO 7000 coolant what used as the cutting fluid. It is manufactured by Blaser Swisslube and it was diluted to a ratio of 1 part oil to 19 parts water and delivered via conventional coolant supply technique at a flow rate of 10 l/min. The fluid concentration was monitored prior each test using an ATAGO N1 refractometer, maintaining the BRIX at approximately 4%. Prior to each grinding test, the grinding wheel was dressed with a single-point diamond dresser, following the conditions described in Table 1, which summarizes the operational conditions employed in this study.

The levels of radial depths of cut (a_e) and workspeeds (V_w) were adjusted in four tests, as shown in Table 2.

After surface grinding wheel preparations and the selection of machining parameters, the grinding tests were performed. Figure 2a–c show the silicon carbide grinding wheel (SiC), the set-up of the grinding experiment with the Polyamide 6 sample mounted in the vise and the procedure for measuring the sample's roughness.

The output variables analyzed in the grinding tests were the roughness parameters Ra (arithmetic mean roughness), Rz (maximum profile height), Rq (root mean square roughness) and Rt (total profile height). For measuring these roughness parameters, a portable surface roughness tester model, Mitutoyo SJ-201P, with a resolution of 0.01 μm and a diamond-tipped stylus with a tip radius of 5 μm was used. A cut-off length of 0.8 mm (sampling length) and an evaluation length of 4.0 mm were adopted. For each test condition, five roughness measurements were taken perpendicular to the grinding wheel's feed marks and the calculation of average and standard deviation of the values were taken based on the experimental results.

Table 1 Grinding conditions

Type of operation	Peripheral surface grinding
Specimens	Polyamide 6 Round shape with dimensions of 50.6 mm (diameter) × 24.4 mm (height)
Grinding wheel	Silicon Carbide (SiC)—39C46KVK 239 mm (outer diameter) × 25.4 mm (width) × 255 mm (inner diameter)
Dressing parameters	Overlap ratio (U_d) = 3 Depth of dressing cut (a_d) = 30 μm Width of the dressing tool (b_d) = 0.3 mm Axial feed (S_d) = 0.1 mm Axial dressing feed rate (V_d) = 240 mm/s Dressing time (t_d) = 6.35 s
Cutting parameters	Cutting speed (V_s) = 30 m/s Workspeeds (V_w) = 2.7–7.5 m/min Radial depths of cut (a_e) = 15 and 45 μm Cutting depth (a_p) = 25.4 mm
Cutting fluid type/delivery technique and flow rate	Vegetable-based semi-synthetic, Vasco 7000, Conventional coolant supply, at a flow rate of 10 l/min, BRIX of 4%

Table 2 Experimental plan designed for the grinding tests

Test number	a_p (mm)	V_s (m/s)	a_e (μm)	V_w (m/min)
1	25.4	30	15	2.7
2				7.5
3			45	2.7
4				7.5

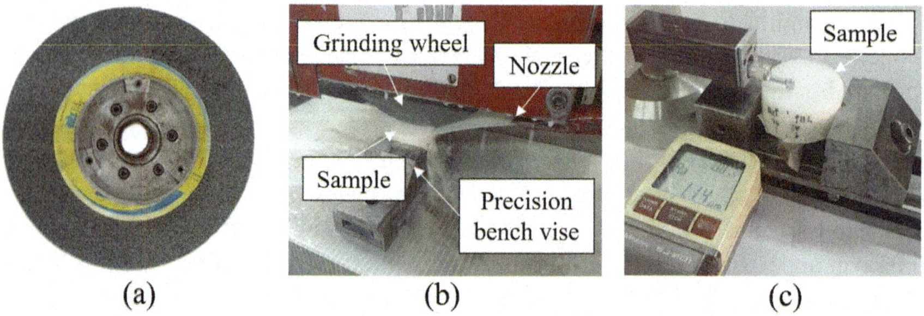

Fig. 2 **a** Silicon carbide (SiC) grinding wheel; **b** grinding experiment set-up; **c** roughness measurement procedure

3 Results and Discussions

Figure 3a–d present the results of Ra, Rz, Rq, and Rt roughness recorded after the peripheral surface grinding tests of Polyamide 6 as a function of radial depth of cut (a_e) and workspeed (V_w).

By analyzing the data shown in Fig. 3a, it can be observed that increasing the radial depth of cut from 15 μm to 45 μm, combined with a workspeed (V_w) of 2.7 m/min, resulted in a 361.29% increase in Ra roughness, while an increase up to 353.54% was observed after machining at $V_w = 7.5$ m/min. According to Klocke [9], this can be attributed to the greater number of abrasive particles in contact with the workpiece, which increases the contact area between the grinding wheel and the workpiece resulting in a higher friction coefficient that causes increased heat generation. This will also require more cutting forces to perform cutting, which changes the tribological condition and adversely affect surface finish. With the increase in radial depth of cut, the grooves left by the abrasive particles on the workpiece surface become deeper, thereby raising surface roughness values and compromising surface quality. These results are similar to those obtained by Spina et al. [10] who performed tangential surface grinding on Polyamide 66 workpieces, varying the radial depth of cut (a_e) from 50 μm to 150 μm while keeping the workspeed (V_w) constant at 1.5 m/min, using a white aluminum oxide grinding wheel (Al_2O_3) with coolant. According

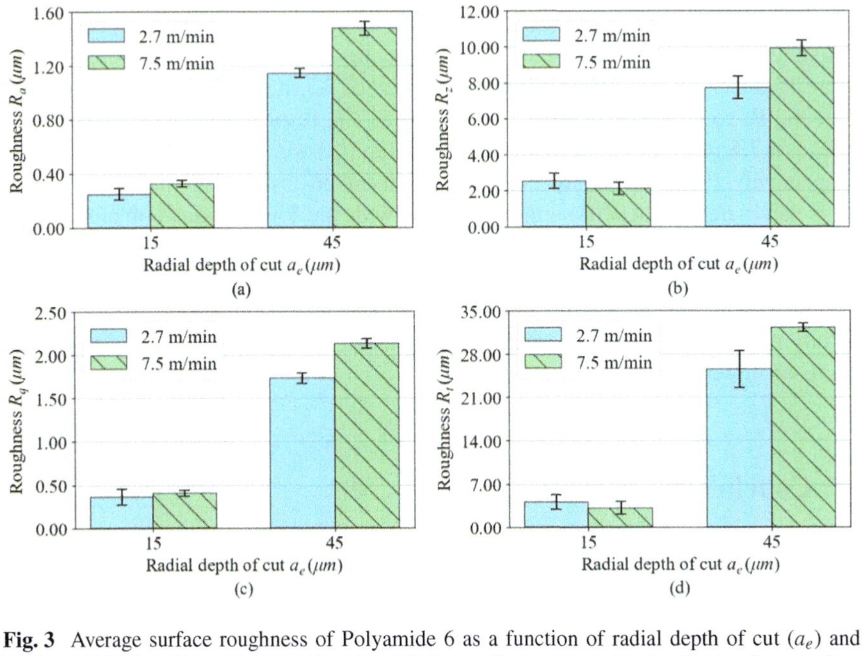

Fig. 3 Average surface roughness of Polyamide 6 as a function of radial depth of cut (a_e) and workspeed (V_w): **a** Ra roughness values (μm); **b** Rz roughness values (μm); **c** Rq roughness values (μm); and **d** Rt roughness values (μm)

to these authors, increasing radial depths of cut during grinding generally results in higher cutting forces and greater plastic deformation on the workpiece surface, which can impair the quality of the machined surface. This occurs due to increases in grinding wheel wear rate that increases friction coefficient because of glazing of abrasive grits generating more heat at the grinding zone and increasing the size of layers with plastic deformation on the workpiece that tends to deteriorate surface finish (high values of surface roughness dominate). Another factor to be considered in these results is the workspeed parameters. Increasing V_w from 2.7 to 7.5 m/min, combined with a radial depth of cut of 15 μm, resulted in a 31.05% increase in Ra roughness. When the radial depth of cut was 45 μm, the change in V_w from 2.7 to 7.5 m/min caused an approximate 28.86% increase in surface roughness. This is due to the increase in material removal rate (Q_w) and equivalent chip thickness (heq), which raises the tensile stress on the abrasive grits and generates higher cutting forces to remove a larger volume of material that results in increased workpiece roughness [11, 12]. This was also found by [6] during the grinding of a Polyamide 66 with aluminum oxide grinding wheel with coolant and workspeed (V_w) varying in range of 0.6 to 1.8 m/min and radial depth of cut (a_e) of 100 μm. In Fig. 3a is noteworthy that even under the most severe tested condition (a_e = 45 μm and V_w = 7.5 m/min), the average Ra roughness of 1.47 μm remained below the rejection limit of Ra = 1.60 μm, as upper limit stipulated for semi finish grinding conditions commonly reported in the literature [2]. This indicates that the range of cutting parameters selected in this study was appropriate for detecting variations in surface roughness of Polyamide 5. Figure 3b–d show a similar trend for the Rz, Rq, and Rt roughness parameters compared to Ra in Fig. 3a. Increasing the radial depth of cut from 15 to 45 μm, combined with a workspeed of 2.7 m/min, led to increases in Rz, Rq, and Rt roughness by 204%, 368%, and 517%, respectively. At a higher workspeed of 7.5 m/min, the increases in Rz, Rq, and Rt were 366%, 280%, and 926%, respectively. Finally, by analyzing Fig. 3a–d, it can be observed all the surface roughness parameters followed a similar pattern, with the lowest values obtained under the conditions of lower a_e and V_w, while increases in workspeed resulted in higher roughness values, with the exception for Rz (Fig. 3b) and Rt (Fig. 3d) parameters when using a radial depth of cut of 15 μm with a variation in workspeed from 2.7 to 7.5 m/min, where a slight reduction in roughness was observed with the associated increase in workspeed.

4 Conclusions

Based on the experimental results obtained from the peripheral surface grinding of Polyamide 6 under different cutting conditions, the following conclusions can be drawn:

- Increasing the radial depth of cut from 15 to 45 μm resulted in a 361.29% increase in Ra roughness with a workspeed of 2.7 m/min, and a 353.54% increase with the workspeed kept constant at 7.5 m/min;
- Increasing the workspeed from 2.7 to 7.5 m/min resulted in a 31.05% increase in Ra roughness with a radial depth of cut of 15 μm and about 28.86% increase with a radial depth of cut of 45 μm;
- Even under the most severe machining conditions ($a_e = 45$ μm and $V_w = 7.5$ m/min), the average Ra roughness was 1.47 μm, which remained below 1.60 μm;
- Increasing the radial depth of cut from 15 to 45 μm, with a workspeed of 2.7 m/min, resulted in increases of 204% in Rz roughness, 368% in Rq roughness, and 517% in Rt roughness, respectively;
- Increasing the radial depth of cut from 15 to 45 μm combined with a workspeed of 7.5 m/min results in increase of 366%, 280% and 926% in Rz, Rq and Rt roughness parameters, respectively.

Acknowledgements The authors thank the Conventional Machining Laboratory (LUC) and Machining Teaching and Research Laboratory (LEPU) and Post-Graduate Program in Mechanical Engineering of the School of Mechanical Engineering at Federal University of Uberlândia. They also thank the Coordination for the Improvement of Higher Education Personnel—CAPES (Brazil), Code de Finança 001, and CNPq via project Process n: 312274/2022-0—Productivity and Research Fellow for financial support for Rosemar Batista da Silva. He is grateful to FAPEMIG via process No. PPM-00265-13, No. APQ-01119-16 and PPM Call 002/2018—Programa Pesquisador Mineiro PPM XII—process No. PPM-00492-18.

References

1. Sawadekar NV, Salunke JJ (2021) An analysis of surface roughness (Ra, Rz), material removal rate and residual stresses in cylindrical grinding of hardened OHNS
2. Ezugwu EO et al (2007) Surface integrity of finished turned Ti–6Al–4V alloy with PCD tools using conventional and high pressure coolant supplies. Int J Mach Tools Manuf 47(6):884–891
3. Groover MP (2014) Introdução aos processos de fabricação, 2nd edn. LTC Livros Técnicos e Científicos Editora SA, Rio de Janeiro
4. Palová K, Kelemenová T, Kelemen M (2023) Measuring procedures for evaluating the surface roughness of machined parts. Appl Sci 13(16):9385
5. Zhang M, Yao Z, Xu S (2014) Optimization of roughness and residual stresses in path-controlled grinding of crankpins. Procedia CIRP 36:117–122
6. Spina R, Cavalcante B (2020) Evaluation of grinding of unfilled and glass fiber reinforced polyamide 6,6. Polymers 12(10):2288
7. Spina R et al (2022) Analysis of surface grinding of thermoplastics specimens with inline measurements. J Manuf Mater Process 6(4):81
8. Hadad M et al (2024) Exploring new parameters to advance surface roughness prediction in grinding processes for the enhancement of automated machining. J Manuf Mater Process 8(1):41
9. Klocke F (2009) Manufacturing processes 2. Springer, Berlin, Heidelberg
10. Spina R, Cavalcante B, Massari M, Rutigliano R (2021) Forces and specific energy of polyamide grinding. Materials 14(17):5041

11. Rowe WB, Andrew W (2014) Principles of modern grinding technology, 2nd edn. Elsevier Inc.
12. Abrão BS, Pereira MF, Da Silva LRR, Machado ÁR, Gelamo RV, De Freitas FMC, Mia M, Da Silva RB (2021) Improvements of the MQL cooling-lubrication condition by the addition of multilayer graphene platelets in peripheral grinding of SAE 52100 steel. Lubricants 9(8):79

Analysis of the Effect of the Threading Head with Axial Compensation on the Axial Force of Advance and Return in the Threading

Raphael Silva Lins, Paulo Sérgio Martins, and Márcio Bacci da Silva

Abstract The manufacturing of internal and external threads ensures precise and reliable assembly of mechanical components. This process is utilized in various applications, including the production of commercial screws for fastening, spindles for movement transmission, micrometric screws for measuring instruments, and screws used in dental and bone implants. In internal threading, several factors can contribute to the premature breakage of taps, such as material ductility, hardness, machine synchronization errors, excessive cutting speed, dimensional and geometric errors in holes, chips at the bottom of the hole, the lubricity of the cutting oil, tool coating, and the type of threading head used. This investigation analyzes the impact of axial compensation threading heads from different manufacturers on the axial forces during the threading of aluminum alloy AA6063. M8 $\times$ 1.25 mm taps, coated with DLC and having a helix angle of 45°, were examined. Machining was performed on a ROMI 1250D CNC machining center with three axes, using two threading heads with different axial compensation from two manufacturers. The tests employed pressurized cutting fluid at 35 kgf.cm^{-2}, with cutting speeds of 40 m.min^{-1} and 60 m.min^{-1}. Axial force measurements were taken on non-passing threaded holes. A complete factorial design with two factors and two levels was used for statistical analysis, with ANOVA to assess the effects of each factor level. Results indicate that the axial threading head influences the axial force in the process, potentially causing premature breakage of the internal threading tool.

Keywords Tap · DLC · Axial force · Axial tapping head · High pressure

R. S. Lins (✉)
Federal Institute of Goiás, Valparaiso de Goiás, GO, Brazil
e-mail: raphael.lins@ifg.edu.br

P. S. Martins
UNA University Center, Belo Horizonte, MG, Brazil

M. B. da Silva
Federal University of Uberlandia, Uberlândia, MG, Brazil

M. C. Gomes et al. (eds.), *Innovations in Brazilian Machining*, Lecture Notes in Mechanical Engineering, https://doi.org/10.1007/978-3-031-92819-2_3

1 Introduction

The 6XXX series aluminum-magnesium-silicon (Al–Mg–Si) alloys are widely used in the aerospace and automotive industries due to their favorable mechanical properties, low density, and high corrosion resistance, coupled with lower energy consumption during machining and forming processes [1].

The microstructure of as-extruded AA6063 alloys in T4 and T6 temper exhibits equiaxed grains with an average size of 94 μm, containing substantial Al–Si–Mn–Fe intermetallic particles formed during solidification. In the T6 temper, needle-shaped precipitates align along the <100> crystallographic directions, ranging in length from 15 to 120 nm, with an average of approximately 35 nm. These precipitates, comprising magnesium and silicon, enhance the alloy's strength, thereby illustrating the significant impact of heat treatment on microstructural characteristics [2]. This material is characterized by high ductility and toughness.

Threaded components are vital in both industrial and daily applications due to their ability to facilitate quick assembly and disassembly. Nuts are the most common element in mechanical assemblies, although internal threads are also employed for clamping. Internal threads primarily experience tensile forces, while screws are subjected to both tensile and torsional forces [3].

In CNC machining, threading heads are employed to increase productivity through higher cutting speeds. Common threading heads include fixed, self-reversing, floating, and axial compensation heads, with the latter designed to address synchronism errors and radial and axial runouts. High-speed steel is predominantly used for taps, sometimes manufactured through powder metallurgy for enhanced durability. Coatings, including diamond-like carbon (DLC), are applied to taps to reduce friction and wear. Bhowmick et al. [4] demonstrated that DLC coatings improved tool life and reduced torque by minimizing aluminum adhesion during the dry machining of Al-Si alloys.

There is a variety of physical vapor deposition (PVD) coatings for TiCN, CrN, and TiB_2 on cemented carbide, Steininger et al. [5] used two types of DLC coatings, with the primary objective of monitoring torque and force. The results showed that the best performance was achieved by employing DLC-coated tools, considering both the measured torque and the analysis of the built-up edge (BUE), built-up layer (BUL) tendency, and chip morphology. Furthermore, the chip morphology produced by the DLC-coated tools exhibited smoother surfaces, straighter edges, and reduced burr formation. Thus, the probability of clogging the pitches and the flutes could be decreased significantly. When drilling aluminum with high-speed steel tools coated with a dense layer of DLC, with around 18% carbon in a diamond structure, there was no significant influence on the roughness of the holes, but tests with dry machining revealed levels of adhesion low cutting speeds (high speed), which can establish itself as an alternative for "green manufacturing" [6].

The different helix angles of taps coated with TiN demonstrated that the torque increased with the cutting speed, while the axial force decreased, using axial compensation heads, Refs. [7, 8] compared threading heads under various cooling conditions, finding that minimal quantity lubrication (MQL) combined with self-reversing heads provided the lowest torque, while emulsion systems with axial compensation yielded the best thrust force results.

Given the lack of research on high-pressure fluid application in threading, especially with aluminum and its alloys, and considering studies like those by Refs. [5, 8] on DLC-coated tools, further investigation into high-speed threading and the efficiency of these coatings is necessary.

2 Materials and Method

AA6063-T6 is a widely used and readily available Al–Si–Mg alloy with low silicon content, sufficient to form magnesium silicates. Table 1 shows the typical composition of this alloy. Primary aluminum alloys are widely used in the automotive sector.

The experiments were carried out at the Mechanical Engineering Laboratory of the UFTM. The equipment used was a CNC Machining Center model 1250D, manufactured by ROMI, with 18 kW of power on the main axis and maximum rotation of 12,000 rpm. Soluble cutting fluid MV AQUA 180, manufactured by VCI, was used at a concentration of 8%. To measure the axial force and torque during operation, a Kistler dynamometer model 9123 C1211 was used, along with a Kistler multi-channel signal conditioner model 5223131. Four taps with a cut coded by the manufacturer as HSS-E M8 × 1.25 mm were used, with a 45° helix angle, with an IKZ-type internal lubrication channel, with a C-type chamfer, manufactured by EMUGE, coated with Diamond-Like-Carbon (DLC), by company Oerlikon Balzers, (see Fig. 1), the pre-holes had an average diameter of 6.81 mm.

The taps were fixed in a threading head with axial compensation, in two different mandrels, called Head 1, model Softsynchro® Modular, manufacturer EMUGE and Head 2 model Synchro-Gewindefutter WN-GM300, manufacturer GUHRING (Fig. 2).

The increase in productivity is directly related to the behavior of the cutting tap when subjected to high cutting speeds. Concomitantly, the axial force compensation

Table 1 Chemical composition of 6XXX series aluminum alloys [9]

Si	Fe	Cu	Mn	Mg	Cr	Zn	Ti	Others		Aluminum (minimum)
								Each	Total	
0.20–0.60	0.35	0.10	0.10	0.45–0.90	0.10	0.10	0.10	0.05	0.15	Remaining

Fig. 1 DLC coated M8 cutting tap with 45° helix angles

Fig. 2 Head with axial compensation, in two different mandrels

Table 2 Cutting parameters and their respective levels

Levels	Factors	
Testes	v_c [m.min^{-1}]	Tapping head
–	40	1
+	60	2

capacity promoted by the tapping head. The synchronization during the advance and return in the internal tapping process, in the CNC Machining Center model 1250D is limited to 2400 RPM, which is the determining factor in the study developed.

A factorial design with 2 levels and 2 factors was used, which are: cutting speed (40 and 60 m.min^{-1}); threading head (type 1 and type 2), with response variable Axial force Fz in advance and return. The works were carried out with the application of cutting fluid at constant high pressure (35 kgf.cm^{-2}), and the helix angle of the tool constant at 45°, as shown in Table 2.

The threading tests were carried out continuously, without breaking strategy of the chip, for a length of 20 mm. Replicates were performed for all conditions. ANOVA analysis of variance was performed, obtaining Pareto charts of standardized effects, interaction graphs and surface graphs to determine whether there is a difference between the means for each cut-off condition. Graphs were also obtained that show the axial forces in the advance and return of the tap as a function of the cutting speed, and the type of axial threading head.

3 Analysis of the Results

The experimental design and methodology described above, it was possible to collect and generate the data presented in this section. Table 3 presents the statistical difference and the effects of the comparisons of the output variables as a function of the input factors (v_c, Tapping head). For *p*-value ≤ 0.05, it means that the input factors influenced the response parameters. The influential factors are highlighted in bold in Table 3. The influence can be visualized through Pareto charts and interaction charts that were analyzed together with the two input factors. Figure 3 demonstrates that comparing the p value with a significance index ($\alpha = 0.05$), the Pareto chart demonstrates the Cutting Speeds (v_c) and Tapping heads (Th) analyzed significantly influence the process, therefore, these two input variables influenced the response surface for the axial buoyancy force (Fz). For v_c equal to 60 $m.min^{-1}$ and for Tapping head type 2, the highest averages of axial thrust force (Fz) were obtained. The Tapping head factor with axial compensation combined with the cutting speed has no influence on the response surface, which can be proven by the Pareto chart that in the joint analysis of the two factors v_c and Th, which can be discarded the joint analysis of these factors, with a *p*-value = 0.095 (*p*-value ≥ 0.05).

This trend was proportional when the type of axial compensation head for fixing the tap cutting machine was varied. For a cutting speed of 40 $m.min^{-1}$ the axial cutting feed force (Fz) had an average variation of 17.8 N, and for a cutting speed of 60 $m.min^{-1}$ the axial feed force of cutting (Fz) had an average variation of 22.3 N. Considering the two threading heads for type 1, the average Fz in the cutting advance was 17 N, with the average values of Fz for the type 2 head of 23.1 N (see Fig. 3).

Table 3 Statistical difference and effect of comparisons between results with input factors

Answers		Cutting speeds (v_c)	Tapping head (Th)	$v_c \times$ Th	S	R2 (%)	R2 (aj) (%)	R2 (pred) (%)
Advance axial force (N)	*p*-value	**0.000**	**0.000**	0.095	2.12783	80.42	76.75	69.41
	Effect	4.550	5.998	1.686				
Axial return force (N)	*p*-value	**0.000**	**0.000**	0.066	5.38207	85.25	80.12	73.84
	Effect	14.23	15.36	4.74				

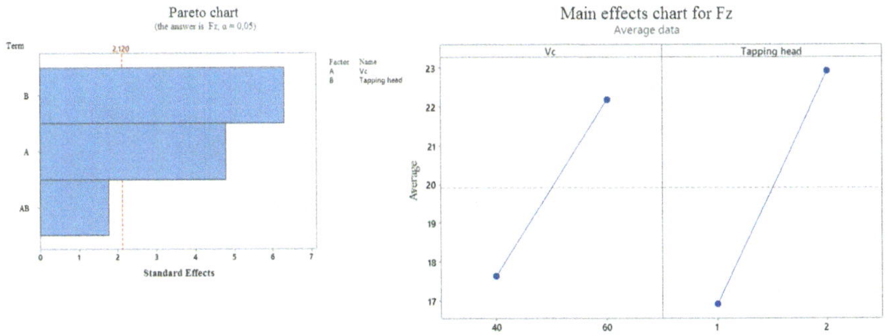

Fig. 3 Pareto chart for Fz (**a**), Interaction chart for means for Fz (**b**)

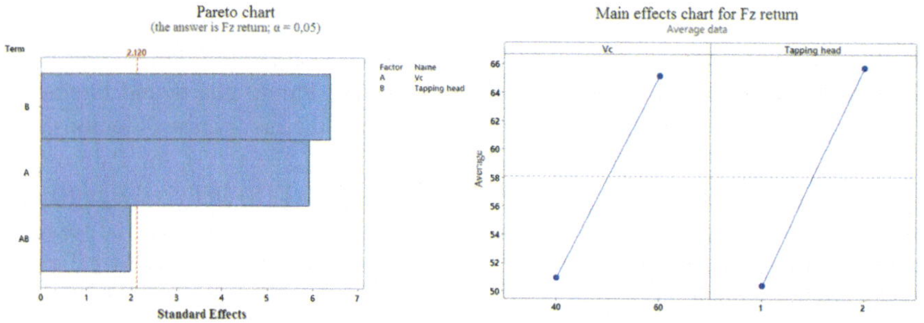

Fig. 4 Pareto chart of standardized effects for Fz return (**a**), Interaction chart for means for Fz return (**b**)

There is a proportional tendency to the return Fz when the type of axial compensation head for fixing the core cutting machine is varied. For a cutting speed of 40 $m.min^{-1}$, the axial cutting feed force upon return of the threading tool had an average variation of 51.2 N, and for a cutting speed of 60 $m.min^{-1}$ Fz return had an average variation of 65.7 N. Considering the two type 1 heads, the average Fz in the cutting return was 50.6 N, with the average Fz return values for the type 2 head of 66 N (see Fig. 4).

Using the surface graph, it is possible to identify that Tapping head type 1 and v_c equal to 40 $m.min^{-1}$ generated the lowest Fz values in the advance and return (e. g. Figure 5a, b). While Tapping head type 2 and v_c equal to 60 $m.min^{-1}$ generated the highest Fz values in the advance and return.

The return axial forces, regardless of the cutting speed used or the threading head used, were greater, due to ductility and the tape chip generated by the 6063-aluminum alloy, which acts as a barrier to the exit of the threading tap, disturbing the return timing and filling the tool clearance surface increasing the friction region. The threading process with an M8 × 1.25 mm tap, coated with DLC, with a cutting speed of 40 $m.min^{-1}$, with a coolant working pressure of 35 $kgf.cm^{-2}$, with a helix angle of 45°, using a type 2 axial threading head, it can be seen that during the cutting

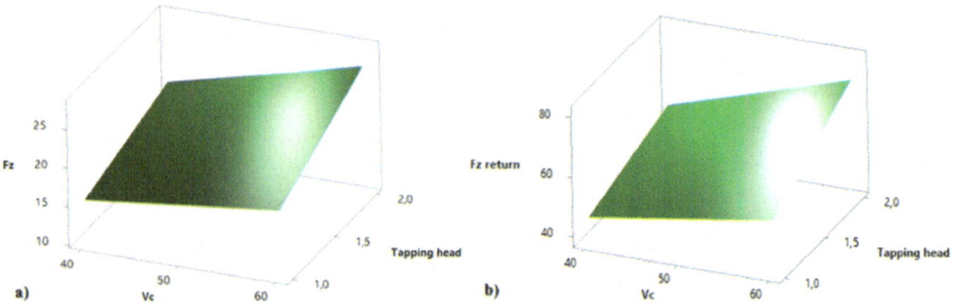

Fig. 5 Surface chart **a** Fz versus v_c versus tapping head; **b** Fz return versus v_c versus tapping head

feed (Fz) the process takes 1.2 s, a programmed pause of 1 s is performed and the return feed occurs in 1.5 s.

Disregarding the coolant pressure in the manufacturing process, since it is the same in advance and return, as well as the helix angle and the cutting advance of 1.25 mm. The cutting speed appears as the main influence on the axial force (Fz) because it is higher in the cutting advance than in the return, because due to the total stop and the resumption of synchronism, the threading tool does not reach the initial cutting speed of the process, due to the cutting length being insufficient to reach 1600 revolutions per minute. It appears that the average axial force amplitudes are greater in the return, which indicates that there is an obstruction caused by the ribbon chip during the return, since there is no cutting force acting on the return of the cutting tap. This requires greater axial compensation of the threading head.

The Fz averages in feed using the tapping head type 2 are 36% higher compared to the tapping head type 1. The averages of Fz return were 2.5 times higher for type 2 compared to type 1. Clearly demonstrating the influence of axial threading head type on force Fz.

A ductility of the AA6063-T6 alloy is the main reason for its poor machinability; the chip does not shear, resulting in a continuous, stringy chip. The application of cutting fluid at a pressure of 35 kgf.cm^{-2} proved to be an excellent alternative for the internal threading process, preventing damage and breakage of the tested cutting taps.

The DLC coating for cutting taps presents itself as a good alternative for machining aluminum alloys using high pressure cutting fluids. The low adhesion on the outlet surface of the taps tested characterizes the use of this coating for moderate and high cutting speeds.

4 Conclusion

According to the results obtained in the experimental tests of high-pressure threading, with an M8 × 1.25 mm cutting tap, coated with DLC, in the AA6063 aluminum alloy, the following conclusions could be obtained from this study.

The axial cutting force at return is three times greater, on average, than the axial force at cutting feed in the experiments carried out, regardless of the cutting speed or axial compensation tapping head in the internal threading processes investigated. This effect is attributed to the continuous chip that is not completely removed even when using a cutting fluid application pressure of 35 kgf.cm^{-2}.

The behavior of the axial force both in the advance and in the return increases with the increase of the cutting speed.

The threading head with axial compensation has the greatest influence on the response surface, being decisive in increasing the axial force of advance and return.

References

1. Zhang Y, Wu Y, Chen D, Wang R, Li D, Guo C, Jiang G, Shen D, Yu S, Nash P (2017) Microstructures and growth mechanisms of plasma electrolytic oxidation coatings on aluminium at different current densities. Surf Coat Technol 321(10):236–246
2. Kula A, Niewczas M (2023) Mechanical properties and rate-sensitive deformation of AA6063 aluminum alloys at 298 K, 78 K, and 4 K. Mater Des 237(11):1–17
3. Brandão GL, Silva PMd, de Freitas SA et al (2020) State of the art on internal thread manufacturing: a review. Int J Adv Manuf Technol 110(3):3445–3465
4. Bhowmick S, Lukitsch MJ, Alpas AT (2010) Tapping of Al-Si alloys with diamond-like carbon coated tools and minimum quantity lubrication. J Mater Process Technol 210(7):2142–2153
5. Steininger A, Siller A, Bleicher F (2015) Investigations regarding process stability aspects in thread tapping Al-Si alloys. Procedia Eng 100(9):1124–1132
6. Ba ECT, Martins PS, Dos Santos CE, Maia LHA, Barbosa JW, Dumont MR (2024) Study of the behavior of skewness (Rsk) and kurtosis (Rku) parameters in the dry drilling process of Al-Mg-Si alloy 6351 T6 using a Diamond-like carbon coated high-speed steel tool. Tribol Int 191(10):1–13
7. Siqueira BS, Freitas SA, Pereira RBD, Brandão LC (2019) Influence of chip breaker and helix angle on cutting efforts in the internal threading process. Int J Adv Manuf Technol 102(9):1537–1546
8. Coelho CCF, Pereira RBD, Lauro CH, Brandão LC (2019) Performance evaluation of tapping processes using a 7075-aluminium alloy with different cooling systems and threading heads. C J Mech Eng Sci 233(19):6793–6806
9. Brazilian Aluminum Association Homepage. http://abal.org.br/aluminio/caracteristicas-quimicas-e-fisicas/. Last accessed 06 Jun 2022

Edge Preparation of Cemented Tungsten Carbide Cutting Inserts with an Abrasive Rotary Brush

Rafaela Silva Barbosa, Ueliton Carvalho Alves, Amauri Hassui, and Carlos Eiji Hirata Ventura

Abstract Due to the high thermomechanical loads in machining processes, an appropriate cutting edge preparation is crucial for achieving longer tool life and good surface finish. Given the high mechanical properties of tool materials, abrasive processes have been used to produce rounded and chamfered edges. Among these processes, brushing stands out due to its flexibility, low cost, and ability to provide efficient edge preparation, allowing the obtaining of specific edge geometries through the correct selection of process parameters. Brushing also offers a controlled and consistent way to optimize the cutting edge, minimizing the risk of defects or excessive wear. However, the correlations between these parameters and the final geometry are associated with the brush-insert contact conditions and must be obtained experimentally. In this study, ground cemented tungsten carbide cutting inserts ISO grade P10 were brushed on an adapted mechanical lathe using a brush with nylon filaments impregnated with SiC abrasives. From the obtained results, it was concluded that penetration depth and inclination angle are the main factors influencing the microgeometry configuration along the main cutting edge and the nose radius. These findings highlight the importance of precise parameter control in optimizing edge preparation for enhanced tool performance.

Keywords Cutting edge preparation · Microgeometries · Brushing · Machining

R. S. Barbosa · C. E. H. Ventura
Department of Mechanical Engineering, Federal University of São Carlos, São Carlos, SP, Brazil

U. C. Alves (✉)
Federal Institute for Education, Science and Technology (Campus Piracicaba), Piracicaba, SP, Brazil
e-mail: ueliton.alves@ifsp.edu.br

A. Hassui
Department of Manufacturing and Materials Engineering, University of Campinas, Campinas, SP, Brazil

M. C. Gomes et al. (eds.), *Innovations in Brazilian Machining*, Lecture Notes in Mechanical Engineering, https://doi.org/10.1007/978-3-031-92819-2_4

1 Introduction

Edge preparation has proven to be a relevant aspect in the study of machining processes. It has the potential to improve the workpiece surface integrity and tool life, as it contributes to avoiding defects caused by previous operations [1]. Therefore, proper edge preparation can increase productivity and reduce process costs. Studies conducted by Denkena et al. [2] demonstrated that the chosen edge preparation process significantly influences the performance of cutting tools, as it impacts surface roughness and the induced residual stresses, which, in turn, have a relevant effect on wear behavior.

Different methods can be used to produce edge rounding with different shapes and sizes [3], which significantly influence the tool wear behavior. Through theoretical and experimental results, Özel [4] observed significant variations in forces, stresses, and temperatures during the machining of hardened steel, due to changes in edge geometry in CBN tools. Ventura et al. [5] also investigated the influence of tool microgeometry on the machinability of hardened steel and noticed that changes in edge geometry can drastically affect forces and temperature, whose increase can damage tool life and the quality of the machined component.

Among the processes used for cutting edge preparation, the most commonly applied, due to the low cost and kinematic simplicity, is brushing, in which a brush with nylon filaments impregnated with abrasive particles rotates against the edge. The edge rounding characteristics are obtained from the variation of the process conditions, such as brush rotation speed, penetration depth, inclination angle, time, cutting material, and abrasive grain size [6].

Drag finishing has also been applied industrially. It consists of the tool's translational and rotational movement in a container with an abrasive medium, which removes tiny amounts of material. Among the parameters that affect microgeometry are movement speeds, processing time, type, and size of the abrasive [7]. Uhlmann et al. [8] achieved good tool life results after using this process in cemented carbide micro-end mills.

Laser processing has been increasingly used due to its high precision and speed, but the investment required has limited its wide application. In addition to the speed of the laser movement over the material to be removed, parameters that determine the emitted energy, such as power and pulse frequency, are essential to define the edge geometry [9]. Moreover, depending on the emitted laser settings and the tool material to be machined, thermally affected zones may occur, compromising the tool's integrity [10].

Finally, research on the use of grinding for edge preparation has been conducted, considering the advantages in terms of productivity and precision. To overcome kinematic limitations and enable the production of edge rounding in the nose radius region of cutting inserts, Denkena et al. [11] developed a rounding discretization method by multiple chamfers. Positive tool life results were obtained through this technique when compared to those generated by commercial tools [12].

Considering the impact of edge preparation on the tool life and surface integrity of a machined component, as well as the simplicity and low cost of the brushing process, which makes it favorable for immediate application, this work focuses on a systematic investigation of the effect of brushing parameters on the shape, size, and finish of edge rounding produced on cemented tungsten carbide cutting inserts. The results are not only academically significant but also have practical implications, as they can guide the immediate application of these findings in the industry, particularly in companies dealing with manufacturing and reconditioning of cutting tools.

2 Material and Methods

Cemented tungsten carbide blanks ISO grade P10 with approximate dimensions 13 mm × 13 mm × 5 mm were finished on a four-axe CNC grinding machine Agathon DOM Plus (maximum rotation speed of 3400 rpm and maximum power of 16 kW) to achieve a geometry SNGN 120412. Grinding was performed with a resin bonded diamond grinding wheel with grain size D46 and concentration C100. The following parameters were applied: cutting speed $v_c = 20$ m/s, axial feed speed $v_{fa} = 4$ mm/min, insert rotation speed to finish the nose radius $v_R = 35°$/s. To avoid the effects of grinding wheel wear, the abrasive tool was conditioned before grinding each blank using a cup-type dresser roll with aluminum oxide abrasives of 180 mesh. For this, a dressing cutting speed $v_{cd} = 10$ m/s and a dressing axial feed speed $v_{fad} = 3$ μm/s were applied.

The edge preparation experiments were conducted on the Imor PRN-320 mechanical lathe, adapted with a device designed to clamp the cutting insert and allow the variation of brushing parameters. A brush with nylon filaments impregnated with SiC 60 mesh abrasives was mounted on a shaft fixed to the three-jaw chuck of the same machine, as shown in Fig. 1. The input parameters used in the brushing process are presented in Table 1. The brush rotation speed and the insert edge angle (angle between the main cutting edge and the brush's rotation axis) were kept constant at $n_b = 475$ rpm and $\chi = 0°$, respectively. The process time, the penetration depth, and the inclination angle were kept constant at $t_b = 30$ s, $a_{pb} = 4$ mm, and $\lambda_b = 0°$, respectively, when they were not the parameters under investigation. Two experiments were performed for each parameter combination, ensuring the reliability of the results. Average values and standard deviations were used for analysis.

As shown in Fig. 2a, the microgeometry is mainly characterized by the edge sections on the rake S_γ and on the clearance S_α faces. The ratio between these values defines the form factor $K = S_\gamma/S_\alpha$, which indicates the tendency of the cutting edge towards the rake ($K > 1$) or the clearance ($K < 1$) face. A symmetrical rounding is indicated by $K = 1$. The parameter Δr defines the smallest distance between the intersection point of the rake and clearance faces (ideally sharp edge) and the actual edge rounding profile. A small Δr denotes a sharp edge, while a high Δr means the rounding resembles a chamfer [11]. The measurement of the edges before and after preparation on the nose radius and the main cutting edge (straight part), as

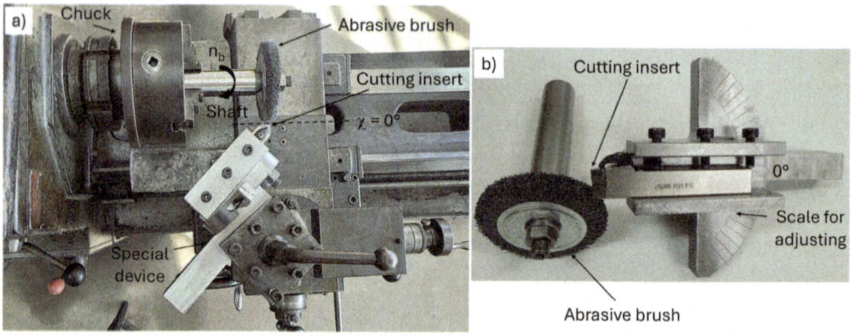

Fig. 1 **a** Top view of the special device mounted on the mechanical lathe and **b** detail of the special device and brush

Table 1 Input parameters

Brushing parameters	Values
Brushing time t_b (s)	30, 60, 120
Penetration depth a_{pb} (mm)	2, 4, 6
Inclination angle λ_b (°)	−20, −10, 0, 10, 20

shown in Fig. 2b, was carried out with a focus-variation microscope Alicona Infinite Focus SL. Two hundred profiles were used to calculate the average values of the mentioned parameters. Additionally, to determine the quality of the prepared edges, the roughness profile was acquired and the maximum roughness parameter Rz was calculated based on five profiles (Fig. 2c).

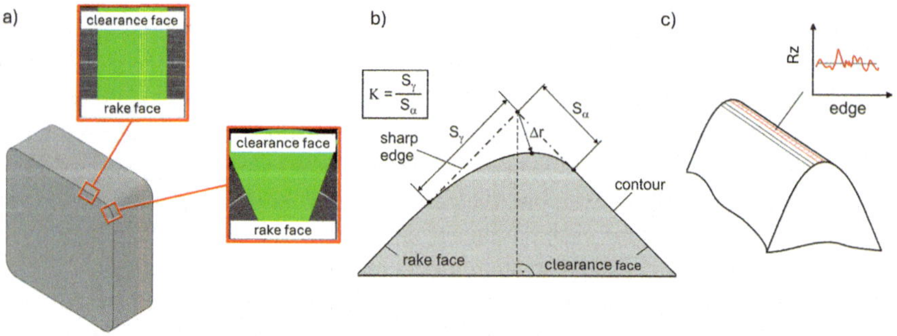

Fig. 2 **a** Characterization of cutting edge geometry, **b** measurement profiles at the straight edge and the nose radius, and **c** measurement of edge roughness

3 Results and Discussion

Representative images of the cutting edge before and after brushing are shown in Fig. 3. The irregularities were characterized by the maximum roughness Rz measured along the straight edge, as indicated in Fig. 2c, and the average values of this parameter before and after the edge preparation experiments are shown in Fig. 4. Considering the standard deviations, it can be observed that different brushing parameters did not lead to significant changes in edge quality. However, a reduction in Rz after brushing was noted in all cases. Before edge preparation, the inserts had an Rz value of approximately 2.5 μm. After brushing, the values ranged from 1.65 to 1.9 μm.

The results obtained after brushing the main cutting edge and the nose radius with different values of penetration depth are demonstrated in Fig. 5a, b, respectively. For the straight edge, an approximately linear increase occurs for $S\alpha$ (from 25 to 70 μm) and $S\gamma$ (from 40 to 90 μm), as well as for Δr (from 12 to 28 μm), indicating that a greater edge rounding can be produced with higher values of penetration depth. In addition, it can be observed that $S\gamma$ presents higher values than $S\alpha$, due to the rotation direction of the brush (from the rake face to the clearance face).

A similar linear trend is observed for the nose radius, but with lower values, possibly due to the less intense contact of the brush with the cutting insert in this region, which also led to the occurrence of coinciding values for $S\alpha$ and $S\gamma$.

Regarding the form factor K (Fig. 5c), defined by the ratio between $S\gamma$ and $S\alpha$, stable values close to 1.3 are observed in the straight edge with the application of different penetration depths. Despite a slight decrease of this parameter in the nose radius region, approximate constant values (between 1.2 and 0.9) are also noted.

The variation of the brushing time was ineffective in altering the investigated parameters, as observed in Fig. 6, where constant values of $S\alpha$, $S\gamma$, Δr, and K can be noted, both for the main cutting edge and the nose radius. This demonstrates that a longer contact time did not promote the removal of a greater volume of material.

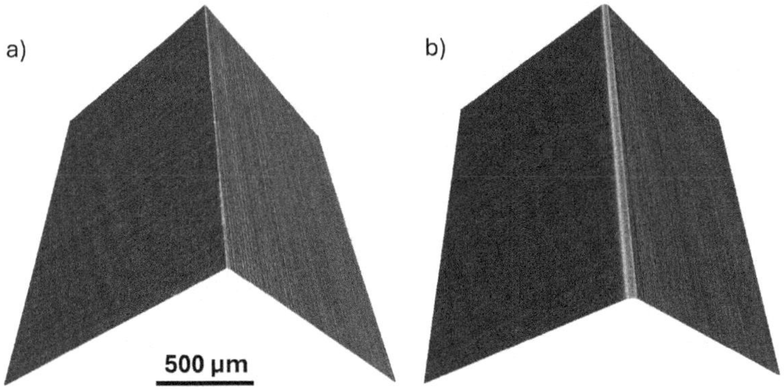

Fig. 3 **a** Sharp edge (after grinding) and **b** brushed edge ($n_b = 475$ rpm, $\chi = 0°$, $\lambda_b = 0°$, $t_b = 30$ s, and $a_{pb} = 4$ mm)

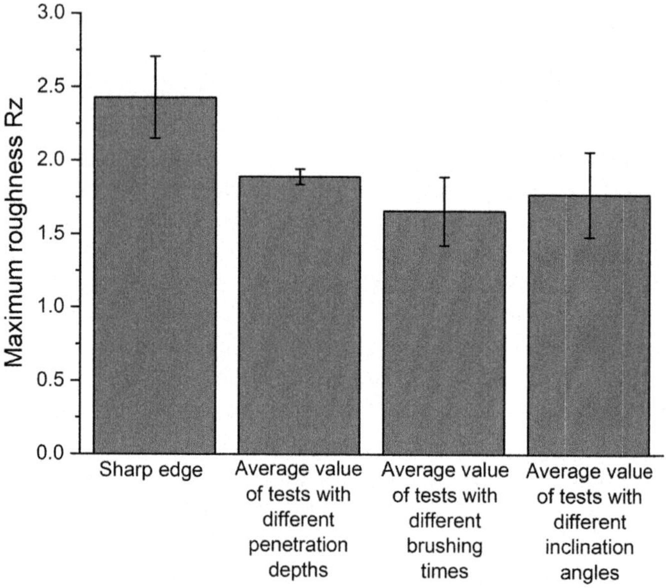

Fig. 4 Maximum roughness Rz along the cutting edge before and after brushing

Within the range of applied parameters, the results show that the processing time should be chosen to complete the process faster. The same explanations given to the results in Fig. 5 apply for the occurrence of higher values of $S\gamma$ at the straight edge (Fig. 6a), which leads to form factors $K > 1$ (Fig. 6c), and similar values of $S\alpha$ and $S\gamma$ at the nose radius (Fig. 6b), which leads to a form factor $K = 1$.

The influence of the inclination angle on $S\alpha$, $S\gamma$, Δr, and K can be observed in Fig. 7. A similar behavior of all output variables is noted for the main cutting edge and the nose radius. Higher inclination angles (more positive) lead to a decrease in $S\alpha$ and Δr and an increase in K, indicating a sharper edge, with asymmetry towards the rake face, due to the brush-insert contact relation. Differently, $S\gamma$ did not present a monotonic behavior. It increased from $\lambda_b = -20°$ to 0° and then decreased from $\lambda_b = 0°$ to $+ 20°$. In the first range, the brush-insert contact area increased with higher inclination angles, while in the second range, higher inclination angles led to the reduction of the filament coverage area. As previously discussed, a minor effect was observed at the nose radius.

The variation of the inclination angle proved to be effective in controlling the form factor K. Form factors $K < 1$ can be achieved using an inclination angle of -20°, both for the nose radius and the main cutting edge. To achieve $K = 1$, an inclination angle of 0° is suitable for the nose radius, while approximately −15° is ideal for the main cutting edge. A form factor $K = 2$ can be obtained on the main cutting edge with an inclination angle of 10° and can be achieved on the nose radius with an angle greater than 20°, considering the trend presented in the graph.

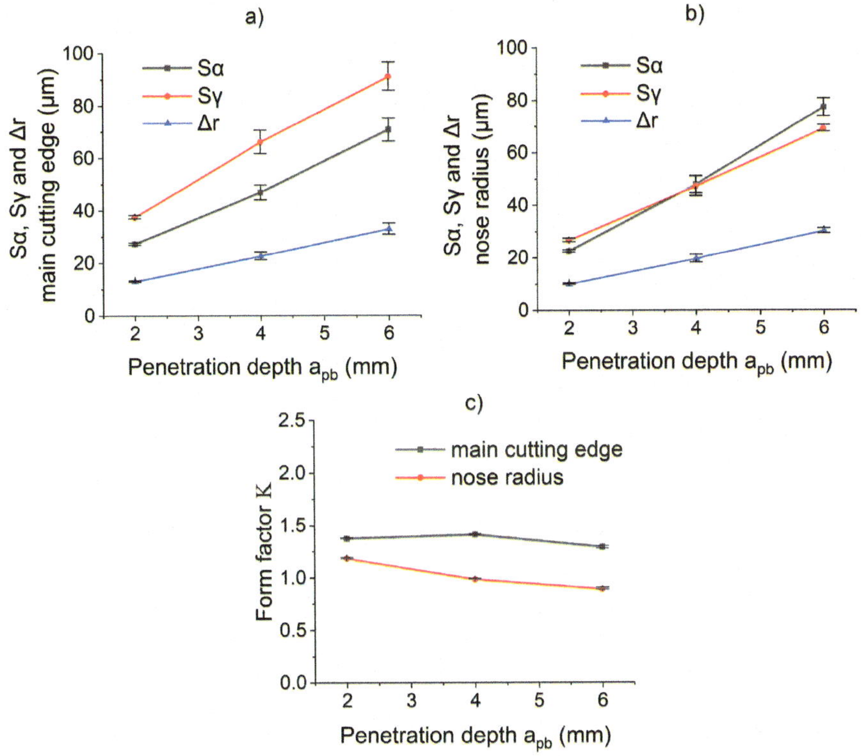

Fig. 5 Behavior of the microgeometry parameters Sα, Sγ, and Δr as a function of the penetration depth at **a** the main cutting edge and **b** nose radius; **c** behavior of the form factor K as a function of the penetration depth at the main cutting edge and nose radius ($n_b = 475$ rpm, $\chi = 0°$, $\lambda_b = 0°$, and $t_b = 30$ s)

Due to the fixed position of the insert in relation to the brush, a regular edge rounding on the nose radius could not be obtained, as observed in Fig. 8. Figure 8a shows a three-dimensional image of a cutting insert brushed with a penetration depth $a_{pb} = 6$ mm, a brushing time tb = 30 s, and an inclination angle $\lambda_b = 0°$, while Fig. 8b, c present the behavior of the microgeometry parameters along the nose radius.

Considering that the main cutting edge was parallel to the brush axis ($\chi = 0°$), greater edge rounding (higher Sα, Sγ, and Δr) was observed along the nose radius towards the main cutting edge, favored by the larger brush-insert contact area. To obtain a regular edge rounding, the penetration depth should be kept constant by contouring the nose radius. This could be done by applying the brushing process on a CNC machine programmed with circular interpolation. Bergs et al. [13] succeeded using a five-axis CNC machine.

Other parameters could also be altered to obtain different values of Sα and Sγ, including the brush's rotation and the insert edge angle. Cross-interactions should

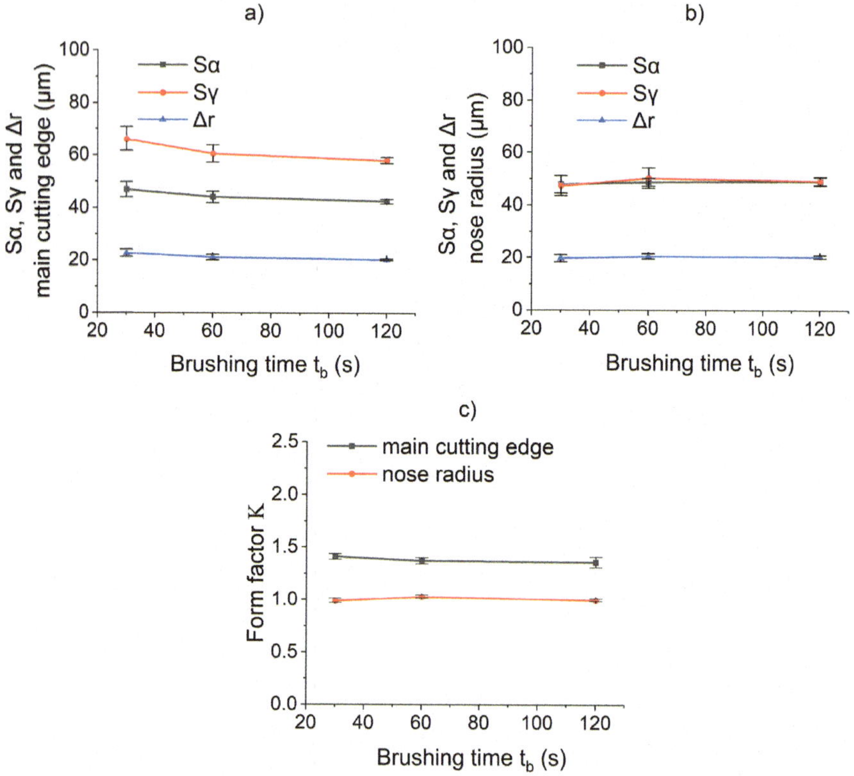

Fig. 6 Behavior of the microgeometry parameters Sα, Sγ, and Δr as a function of the brushing time at **a** the main cutting edge and **b** nose radius; **c** behavior of the form factor K as a function of the brushing time at the main cutting edge and nose radius ($n_b = 475$ rpm, $\chi = 0°$, $\lambda_b = 0°$, and $a_{pb} =$ 4 mm)

also be analyzed to determine the trends more accurately. Such aspects should be addressed in future work.

4 Conclusions

Based on the results obtained from the brushing tests performed on cemented tungsten carbide inserts, the application potential of the method is evident. It can be concluded that:

- The brushing process can reduce the irregularities of the edge rounding and promote a lower edge roughness.
- Higher values of penetration depth lead to a linear increase in the values of Sα, Sγ, and Δr.

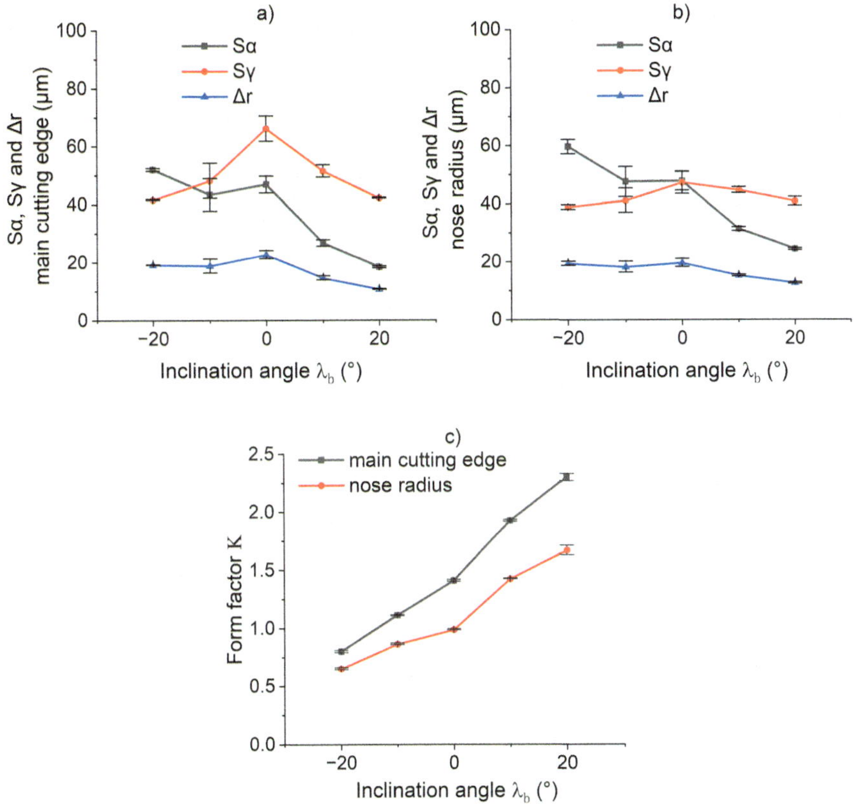

Fig. 7 Behavior of the microgeometry parameters Sα, Sγ, and Δr as a function of the brushing time at **a** the main cutting edge; **b** nose radius; and **c** behavior of the form factor K as a function of the inclination angle at the main cutting edge and nose radius ($n_b = 475$ rpm, $\chi = 0°$, $t_b = 30$ s, and $a_{pb} = 4$ mm)

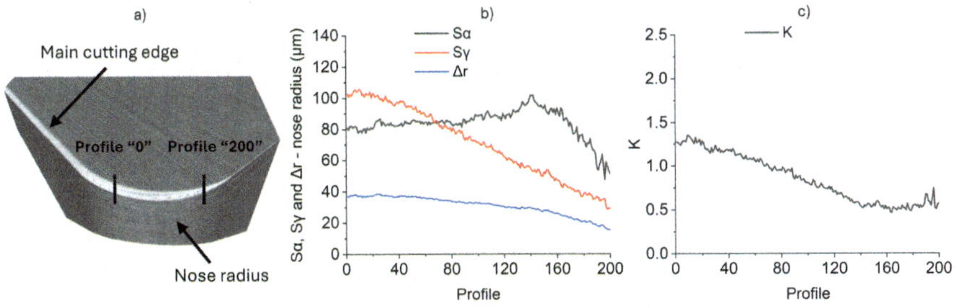

Fig. 8 **a** Three-dimensional image of a cutting insert after brushing and the behavior of the microgeometry parameters **b** Sα, Sγ, Δr, and **c** K along the nose radius

- Changes in the form factor K are closely related to the inclination angle. Values between 0.5 and 2 could be achieved within the range of studied parameters ($-20° < \lambda_b < +20°$).
- Increasing the brushing time does not alter the microgeometry. It only reduces productivity.
- With the proposed adapted process, a constant edge rounding cannot be achieved along the nose radius, significant variations being observed mainly for $S\gamma$.

Acknowledgements The authors acknowledge the São Paulo Research Foundation (FAPESP) [grant numbers 2015/15622-2, 2017/12309-7, and 2017/12304-5], the Coordenação de Aperfeiçoamento de Pessoal de Nível Superior (CAPES) [grant number Finance Code 001], and the National Council for Scientific and Technological Development (CNPq) [grant number 305730/2022-3] for financial support.

References

1. Peter P, Boris P, Tomás V, Jozef P, Marek V, Vladimír S (2020) Cutting edge radius preparation. In: 2018 2nd International conference on nanomaterials and biomaterials on materials today: proceedings, vol 22, pp 212–218
2. Denkena B, Köhler J, Breidenstein B, Abrão AM, Ventura CEH (2014) Influence of the cutting edge preparation method on characteristics and performance of PVD coated carbide inserts in hard turning. Surf Coat Technol 254:447–454
3. Karpat Y, Özel T (2008) Mechanics of high speed cutting with curvilinear edge tools. Int J Mach Tools Manuf 48:195–208
4. Özel T (2003) Modeling of hard part machining: effect of insert edge preparation in CBN cutting tools. J Mater Process Technol 141(2):284–293
5. Ventura CEH, Chaves HS, Rubio JCC, Abrão AM, Denkena B, Breidenstein B (2017) The influence of the cutting tool microgeometry on the machinability of hardened AISI 4140 steel. Int J Adv Manuf Technol 90:2557–2565
6. Denkena B, Leon L, Bassett E, Rehe M (2010) Cutting edge preparation using abrasive brushing. Key Eng Mater 438:1–7
7. Gegenheimer H (2009) Präparationsverfahren: Gleitschleppschleifen. In: TIKAL F (Org.) Schneidkantenpräparation: Ziele, Verfahren und Messmethoden. Kassel University Press, Kassel
8. Uhlmann E, Oberschmidt D, Löwenstein A, Kuche Y (2016) Influence of cutting edge preparation on the performance of micro milling tools. In: 7th HPC 2016—CIRP conference on high performance cutting on Procedia CIRP, vol 46, pp 214–217
9. Aurich JC, Zimmermann M, Leitz L (2011) The preparation of cutting edges using a marking laser. Prod Eng Res Dev 5:17–24
10. Lütjens P (2011) Mit kalter Ablation zur optimalen Schneidkante. Werkstatt und Betrieb 9:132–135
11. Denkena B, Köhler J, Ventura CEH (2013) Customized cutting edge preparation by means of grinding. Precis Eng 37(3):590–598
12. Ventura CEH, Köhler J, Denkena B (2013) Cutting edge preparation of PCBN inserts by means of grinding and its application in hard turning. CIRP J Manuf Sci Technol 6:246–253
13. Bergs T, Schneider SAM, Amara M, Ganser P (2016) Preparation of symmetrical and asymmetrical cutting edges on solid cutting tools using brushing tools with filament-integrated diamond grits. In: 53rd CIRP conference on manufacturing systems on Procedia CIRP, vol 93

Evaluation of Ground Surface Quality of 316L Stainless Steel Produced by Additive Manufacturing

Marcus Vinícius de Oliveira, Josenilton dos Santos Lopes, Alcione dos Reis, Mayara Fernanda Pereira, Louriel Oliveira Vilarinho, Diandro Bailoni Fernandes, Mark James Jackson, and Rosemar Batista da Silva

Abstract Additive manufacturing can allow production of many components that includes complex geometries. However, achieving surface quality comparable to those obtained through turning and milling, for example, still becomes challenging. These challenges can prevent the immediate use of components manufactured through additive manufacturing in operational applications. Therefore, post-processing techniques to enhance surface quality are necessary. In general, a machining operation is required as the first step after producing an additively manufactured part and, depending on the project requirements, the use of grinding processes may be an alternative to post-processing of parts produced by additive manufacturing. Grinding can provide a combination of narrow dimensional tolerances with low surface roughness values (Ra $\leq$ 1.6 μm) as well as allowing the machining of medium to high hardness materials that are limited by machining processes using defined tool geometry. In this context, this work aims to evaluate the roughness of a 316L stainless steel grade manufactured by additively manufacturing using the arc deposition process (WAAM) after grinding with a white aluminum oxide grinding wheel (AA46K6V), which is typically used for grinding steels produced by conventional processes. The cutting conditions employed were radial depth of cut (a_e) values of 5 μm and 35 μm and workspeed (v_w) values of 2.7 m/min and 7.5 m/min, respectively. Ra and Rz roughness parameters were used as output variables. The results showed that the combination between the highest radial depth of cut with the lowest workspeed resulted in a 113% increase in roughness compared to the value at $a_e = 5$ μm. Furthermore, by observing SEM micrograph images, it was observed that the surface finish deteriorated after grinding with the highest radial depth of cut

M. V. de Oliveira (✉) · J. dos Santos Lopes · A. dos Reis · M. F. Pereira · L. O. Vilarinho · D. B. Fernandes · R. B. da Silva
Federal University of Uberlandia, Uberlândia, Minas Gerais, Brazil
e-mail: marcus.vidol@ufu.br

M. J. Jackson
College of Technology and Aviation, Aerospace and Technology Campus, Kansas State University, Salina, KS, USA

M. C. Gomes et al. (eds.), *Innovations in Brazilian Machining*, Lecture Notes in Mechanical Engineering, https://doi.org/10.1007/978-3-031-92819-2_5

($a_e = 35$ μm), thereby resulting in less uniform grooves compared to the workpieces ground with the smallest radial depth of cut ($a_e = 5$ μm).

Keywords Grinding · Additive manufacturing · 316L stainless steel · Radial depth of cut · Roughness

1 Introduction

Additive manufacturing (AM), commonly known as 3D metal printing, is a manufacturing process in which materials are deposited, typically layer-by-layer, according to a predefined three-dimensional model. This process is defined by the ISO/ASTM 52900 Standard [1] with the advantage of producing parts close to their final shape and dimensions, as it involves material addition rather than mechanical removal, which is common in conventional machining processes. However, there is currently no AM process capable of producing a part in its final dimension only through material addition without the need for post-processing. Generally, post-processing includes roughing and/or finishing operations, which can be performed by turning, milling and grinding, depending on the dimensional and geometric tolerances specified in the project.

Grinding is a widely used post-processing technique used in the industry to enhance the surface quality of parts, providing tighter dimensional deviations with low roughness values (Ra ≤ 1.60 μm) [2]. The efficiency of the grinding process depends, among other factors, on the workpiece material, the type and properties of grinding wheel, as well as the correct selection of major cutting parameters such as cutting speed, workspeed, radial depth of cut and the wheel conditioning. In the case of materials produced by AM, such as stainless steel manufactured by Powder Bed Fusion (PBF), surface integrity can be compromised by defects such as spatter and lack of fusion [3, 4]. In general, components manufactured by AM present some defects due to complex melt pool dynamics and solidification microstructures, such as porosity or lack-of-fusion, which size and density will depend on the AM process [5]. Presence of pores can reduce tensile strength and ductility by acting as stress concentrators, thereby affecting their thermo-mechanical properties as well as reducing the grindability of such components [5]. Research is ongoing to investigate and improve the surface quality of these materials after grinding [3, 6, 7]. Typically, parts made by AM are initially subjected to turning and milling processes before undergoing grinding.

Research on the grinding of metallic materials produced by PBF has shown promising results in terms of low roughness values (Ra between 0.30 and 0.90 μm). Brinksmeier et al. [8] achieved these results using a radial depth of cut (a_e) of 10 μm, a workspeed (v_w) of 2.5 m/min and a cutting speed (V_s) of 20 m/s, with a vitrified corundum grinding wheel under dry grinding condition. Liu et al. [9] obtained similar results with $a_e = 20$ μm, $v_w = 1.5$ m/min and $V_s = 25$ m/s using a corundum wheel in presence of coolant.

Grinding removes the surface layer of unmelted powder and substantially reduces surface roughness as well as improve fatigue performance of component [10]. However, there is no direct correlation between surface roughness and fatigue life, possibly due to changes in material properties after grinding [11]. The fatigue performance of metal materials manufactured by AM and ground can be influenced by four workpiece parameters that includes surface finish, porosity, microstructure and residual stress [4]. In this context, this work aims to evaluate the roughness (Ra and Rz parameters) of 316L stainless steel produced by Wire Arc Additive Manufacturing (WAAM) after peripheral surface grinding with a white aluminum oxide (Al_2O_3) grinding wheel. Radial depth of cut (a_e) and workspeed (v_w) were the input variables tested. Surface roughness and surface of workpiece material were examined via Scanning Electron Microscopy (SEM) to analyze the texture of the samples and identify the presence or absence of events related to thermal damage, such as burning or cracks, for example.

2 Methodology

The grinding tests were performed on a MELLO semi-automatic peripheral surface grinder, model P-36, with a constant speed rotation of 2400 rpm, a power of 2.25 kW and resolution of 5 μm on the axis which moves the grinding wheel spindle in a vertical direction. A white aluminum oxide (Al_2O_3) grinding wheel with designation AA46K6V was used, with dimensions of 255.0 mm (outer diameter) × 25.4 mm (width) × 127.0 mm (inner diameter). The material used was an AISI 316L austenitic stainless steel alloy, manufactured by WAAM. The workpieces in rectangular bar of 200 mm (length), 20 mm (width) and 100 mm (height) (pre-form) were previously subjected to a water jet machining process for preparation. After that, the samples were subjected to milling process and then resulted in final dimensions of 95 mm (length), 23 mm (width) and 11 mm (height) (Fig. 1).

In Fig. 1 can also be observed that the material deposition direction used in AM processes is the same as the grinding wheel feed direction used in the grinding tests.

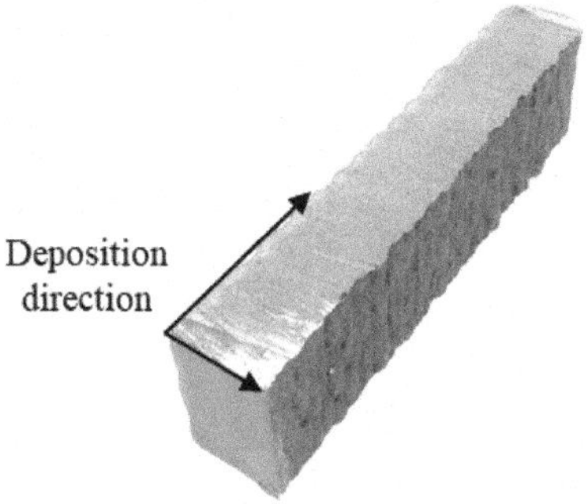

Fig. 1 Workpiece material: AISI 316L austenitic stainless steel manufactured by WAAM

A semi-synthetic emulsifiable type, VASCO 7000, from the Blaser Swisslube manufacturer, was used as cutting fluid, in in the proportion of 1 part of oil to 19 parts of water and delivered by the conventional method at a flow rate of 10 l/min. Its concentration was monitored after each test considering the BRIX of approximately 4% with aid of a ATAGO N1 refractometer. Prior to each grinding test, the grinding wheel was dressed according to the conditions described in Table 1, which summarizes the operational conditions used in this work. The images of the grinding wheel used, the grinding wheel dressing system set-up and the roughness measurement system setup are shown in Fig. 2a–c, respectively.

The roughness parameters Ra and Rz were the output variables analyzed after the grinding tests with aid of a portable surface roughness tester model SJ201P (from Mitutoyo manufacturer) with a resolution of 0.01 μm. The probe has a diamond tip and has a tip radius of 5 μm. A cut-off (sampling length) of 0.8 mm and an

Table 1 Experimental conditions

Grinding operation type	• Peripheral surface
Workpiece (hardness)/dimensions	• AISI 316L austenitic stainless steel (WAAM)—(315 HV) • 95 mm × 23 mm × 11 mm
Grinding wheel	• White aluminum oxide (Al_2O_3)—AA46K6V
Input parameters	• Cutting speed (V_s) = 32 m/s • Workspeed (v_w) = 2.7–7.5 m/min • Radial depth of cut (a_e) = 5–35 μm • Width of cut (a_p) = 23 mm
Dressing parameters	• Dressing overlap ratio (U_d) = 3 • Dressing depth (a_d) = 30 μm (three passes) • Dressing active width (b_d) = 0.3 mm • Dressing lead (S_d) = 0.1 mm • Dressing speed (V_d) = 240 mm/min
Cutting fluid/delivery method	• Semi-synthetic vegetable based-Vasco 7000 • Conventional at flow rate of 10 l/min and dilution of oil in water 1:19

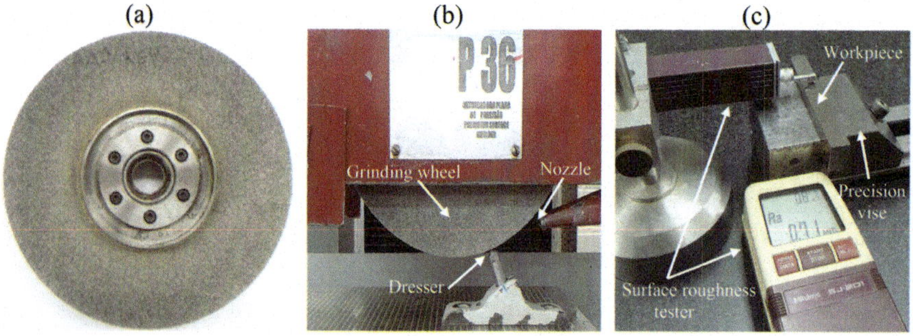

Fig. 2 **a** White aluminum oxide grinding wheel (AA46K6V); **b** grinding wheel dressing; **c** workpiece roughness measurement apparatus

evaluation length of 4.0 mm were adopted for all measurements. For each workpiece, five roughness measurements were taken perpendicular to the feed marks left by the abrasive particles of the grinding wheel. Since the surface roughness tester used lacks a calibration certificate, measurement uncertainty for the roughness parameters Ra and Rz was calculated. To address this, five input variables were analyzed: the average of values indicated by the roughness tester considering the five measurement cycles performed ($\bar{L}$), corrections associated with the resolution of the roughness tester (ΔR), corrections associated with the uncertainty of the roughness tester calibration (ΔC), corrections associated with the stylus tip radius (ΔTr) and corrections due to the sample deformation (ΔD) [12]. Additionally, the workpiece was removed from the precision vise and cleaned with isopropyl alcohol in an ultrasonic cleaner for 10 min. In the sequence, it was dried with compressed air before undergoing the analysis of the machined surfaces was conducted using a SEM, from Hitachi manufacturer, model TM3000, to evaluate the texture of the ground surfaces.

3 Experimental Results and Discussions

This session presents the results of roughness (Ra and Rz) and SEM images of ground surfaces.

3.1 Surface Roughness

The surface roughness results, Ra and Rz parameters, recorded for the 316L stainless steel produced by WAAM after grinding are shown in Fig. 3a, b, respectively. Additionally, error bars indicating the expanded uncertainty are also shown.

Upon analyzing the data from Fig. 3a, it can be observed that increasing the radial depth of cut from 5 to 35 μm, combined with a workspeed of 2.7 m/min, resulted in a 113% increase in roughness. By keeping the workspeed constant at 7.5 m/min an increase by about 100% was observed. According to Klocke et al. [13], this relationship can be attributed to the higher number of abrasive particles in contact with the workpiece, which consequently increases the contact area between the grinding wheel and the workpiece, thereby leading to greater friction coefficient what causes higher heat generation and cutting forces to be elevated. These changes worsen the tribological conditions, negatively affecting the surface finish. The greater the radial depth of cut, the deeper the grooves left by the abrasive particles on the workpiece surface, leading to higher roughness values and negatively impacting the surface quality of the part. According to De Paiva et al. [14], increasing the radial depth of cut results in more intense plastic deformation of the workpiece, which in turn intensifies heat in the grinding zone. This increased temperature on the workpiece surface leads to greater thermal expansion and displacement of the outer surface towards the grinding wheel, which tends to generate high temperatures, resulting in

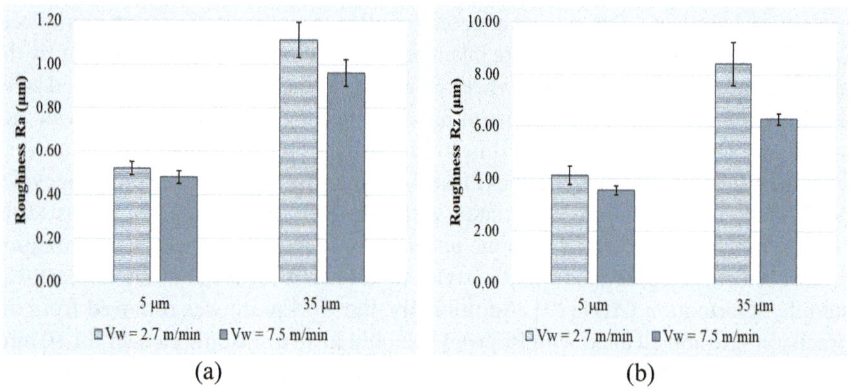

Fig. 3 Results of the roughness parameters Ra (**a**) and Rz (**b**) recorded on the 316L stainless steel produced by WAAM in the grinding tests as function of radial depth of cut (a_e) and workspeed (v_w)

the production of low-quality surfaces characterized by increased surface roughness values.

Another relevant fact to be considered from results in Fig. 3 is the increase in roughness values when reducing the workspeed (v_w) from 7.5 to 2.7 m/min, combined with the highest radial depth of cut (35 μm), which resulted in 14% increase in Ra parameter. According to the literature, this occurs due to the prolonged contact time between the workpiece and the grinding wheel, especially when machining with greater radial depths of cut values, which results in unfavorable tribological conditions, thereby deteriorating the surface finish. In a recent study carried out by Junior [15], the authors observed similar results when grinding the same workpiece material used in this study, but with variations in radial depth of cut in range of 10 μm to 30 μm with workspeed values of 2.7 m/min and 7.5 m/min using a white aluminum oxide grinding wheel (65A120K8V).

By examining the Rz roughness parameter in Fig. 3b, a similar pattern to that observed for the Ra roughness parameter (Fig. 3a) was observed, i.e., increase in Rz roughness with radial depth of cut value. The highest roughness values were recorded when machining with highest radial depth of cut $a_e = 35$ μm combined with the workspeed v_w of 2.7 m/min, resulting in a 103% increase in Rz when varying the radial depth of cut from 5 to 35 μm and keeping the workspeed constant at 2.7 m/min.

In this study is important to highlight, especially concerning the Ra parameter, which is widely used to assess machined surfaces, that it remained below 1.20 μm, which is well below the reference limit for Ra roughness, established at 1.60 μm for semi-finishing grinding conditions [16]. Additionally, a significant improvement in roughness values, especially for Ra and Rz, is observed when using the lowest radial depth of cut, a_e, of 5 μm.

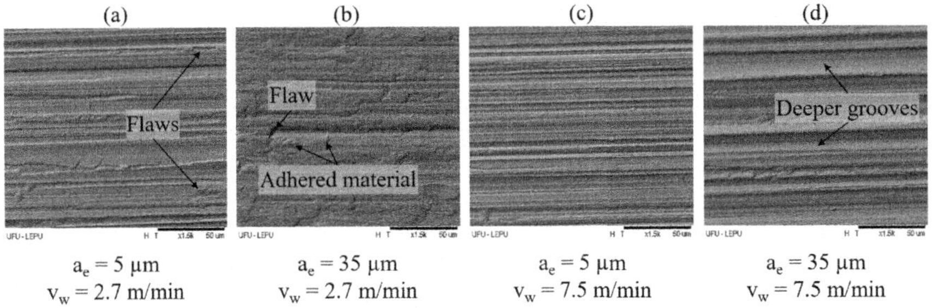

Fig. 4 The SEM images of 316L stainless steel produced by WAAM surfaces after grinding at various cutting conditions

3.2 SEM Images of Ground Surfaces

In Fig. 4a–d are shown the SEM images of the WAAM workpieces surfaces obtained after grinding under different cutting conditions.

From the images in Fig. 4b, d can be observed that the surface finish was compromised when using larger radials depths of cut (35 μm), which marks or grooves are less uniform compared to parts ground with a lowest depth of cut ae of 5 μm (Fig. 4a, c). Owing to the increased contact area between the abrasive particles from grinding wheel and the workpiece surface resulting from the greatest depth of cut value, there is a higher occurrence of plastic deformation during chip formation, which affects the tribological conditions of the process. This leads to an elevated temperature in the cutting region, negatively impacting the surface finish [17]. In a study carried out by Guo et al. [18], the authors performed grinding the of a CoCrFeMnNi high-entropy alloy produced by Selective Laser Melting (SLM) and reported deformed material and deep grooves on the ground surfaces, which was responsible for deteriorating the surface finish of the parts.

Additionally, the occurrence of flaws was more pronounced on surfaces shown in Fig. 4a, b, which can be attributed to higher heat generation usually occurring when machining at a low workspeed ($v_W = 2.7$ m/min), where the contact time between the abrasive grains and the workpiece is prolonged. This heat being concentrated to the heating zone after some time is majorly transferred to the workpiece when machining with conventional grinding wheel abrasive. This will leads to more intense plastic deformation, material redeposition and/or adhered material to the workpiece surface [2]. In general, under these conditions, the grinding wheel is more prone to clogging, which leads to an increase in both normal and tangential cutting forces.

4 Conclusions

After grinding the 316L stainless steel produced by WAAM, the following conclusions can be drawn:

- Increasing the radial depth of cut from 5 to 35 μm combined with a workspeed of 2.7 m/min, resulted in a 113% increase in Ra. The same increase in a_e combined with a workspeed of 7.5 m/min resulted in a 100% increase in Ra;
- Reducing the workspeed (v_w) from 7.5 to 2.7 m/min, combined with the highest radial depth of cut of 35 μm, resulted in a 14% increase in Ra;
- For the Rz parameter, the highest values resulted in a 103% increase when using the highest radial depth of cut (35 μm) combined with the lowest workspeed (2.7 m/min);
- The Ra parameter remained below 1.20 μm, which is well below the upper reference limit set at 1.6 μm for semi-finishing grinding process;
- Surface finish becomes worse when a higher depth of cut value of 35 μm was used, resulting in less uniform grooves compared to workpieces ground with a smaller radial depth of cut ($a_e = 5$ μm), as observed from the SEM images of the ground surfaces.

Acknowledgements The authors thank the Traditional Machining Laboratory (LUC) and Machining Teaching and Research Laboratory (LEPU) and Post-Graduate Program in Mechanical Engineering of the School of Mechanical Engineering at Federal University of Uberlândia. They also thank the Coordenação de Aperfeiçoamento de Pessoal de Nível Superior (CAPES) (Brazil), Code de Finança 001. Rosemar Batista da Silva also thanks to CNPq via project Process n: 312274/2022-0—Productivity and Research Fellow for financial support and to FAPEMIG via Process No. PPM-00265-13, No. APQ-01119-16 and PPM Call 002/2018—Programa Pesquisador Mineiro PPM XII—Process No. PPM-00492-18.

References

1. ISO/ASTM F52910 (2018) Additive manufacturing—design—requirements, guide-lines and recommendations, 1st edn. ISO/ASTM International, West Conshohocken
2. Malkin S, Guo C (2008) Grinding technology: theory and application of machining with abrasives. Industrial Press Inc.
3. Maleki E, Bagherifard S, Bandini M, Guagliano M (2021) Surface post-treatments for metal additive manufacturing: progress, challenges, and opportunities. Addit Manuf 37:101619
4. Zhang H, Li C, Yao G, Shi Y, Zhang Y (2022) Investigation of surface quality, microstructure, deformation mechanism, and fatigue performance of additively manufactured 304L stainless steel using grinding. Int J Fatigue 160:106838
5. Malakizadi A, Mallipeddi D, Dadbakhsh S, M'Saoubi R, Krajnik P (2022) Post-processing of additively manufactured metallic alloys–a review. Int J Mach Tools Manuf 179:103908
6. Zhang H, Xu M, Kumar P, Li C, Liu Z, Zhang Y (2021) Fatigue life prediction model and entropy generation of 304L stainless steel fabricated by selective laser melting. J Mater Process Technol 297:117279

7. Sanaei N, Fatemi A (2021) Defects in additive manufactured metals and their effect on fatigue performance: a state-of-the-art review. Prog Mater Sci 117:100724
8. Brinksmeier E, Levy G, Meyer D, Spierings AB (2010) Surface integrity of selective-laser-melted components. CIRP Ann 59:601–606
9. Liu Z, Li X, Wang X, Tian C, Wang L (2019) Comparative investigation on grindability of Inconel 718 made by selective laser melting (SLM) and casting. Int J Adv Manuf Technol 100:3155–3166
10. Yu H, Li F, Wang Z, Zeng X (2019) Fatigue performances of selective laser melted Ti-6Al-4V alloy: influence of surface finishing, hot isostatic pressing and heat treatments. Int J Fatigue 120:175–183
11. Bagehorn S, Wehr J, Maier HJ (2017) Application of mechanical surface finishing processes for roughness reduction and fatigue improvement of additively manufactured Ti-6Al-4V parts. Int J Fatigue 102:135–142
12. Souza CC, Arantes LJ, Piratelli-Filho A, Arencibia RV (2019) Assessment of the effect of stylus tip radius on milled, bored, and honed surfaces. Int J Adv Manuf Technol 104:2459–2471
13. Klocke F, Fritz Klocke E, Kuchie A (2009) Grinding. In: Manufacturing processes 2: grinding, honing, lapping, pp 1–166
14. De Paiva RL, Da Silva RB, Jackson MJ, Abrão AM (2017) The influence of cutting fluid concentration on surface integrity of VP80 steel and the influence of cutting fluid flow rate on surface roughness of VPATLAS steel after grinding. ASME J Manuf Sci Eng 139:121003
15. Junior FRF (2022) Influência de parâmetros de corte na retificação do aço inoxidável AISI 316L produzido por manufatura aditiva [Influence of the cutting parameters on grinding of AISI 316L stainless steel manufactured by additive manufacturing process] (in Portuguese). Projeto de Fim de Curso (Graduação)—Universidade Federal de Uberlândia, Uberlândia, p 78
16. Diniz AE, Marcondes FC, Coppini NL (2014) Tecnologia da usinagem dos materiais. Artliber [Machining technology of materials] (in Portuguese)
17. Guimarães C (2016) Retificação plana de aços para moldes e matrizes em várias condições de corte e diferentes técnicas de aplicação de fluido de corte [Integrity surface study of mold and die steels after peripheral surface grinding under cutting conditions and different cooling delivery techniques] (in Portuguese). Tese (Doutorado)—Universidade Federal de Uberlândia, Uberlândia, p 189
18. Guo J, Goh M, Zhu Z, Lee X, Nai MLS, Wei J (2018) On the machining of selective laser melting CoCrFeMnNi high-entropy alloy. Mater Des 153:211–220

Influence of Feed on Wear, Electric Current, and Chip Morphology in the Drilling of SAE 1020 Steel

Lívia Letícia da Silva Araújo, Patrick Abreu de Oliveira, Marcus Vinicius da Silva Mendes, Renan Matos Monção, and Marcos Guilherme Carvalho Braulio Barbosa

Abstract Drilling is one of the most widely used and technically demanding machining processes in the industrial sector. Effective control of chip formation is crucial for balancing cutting tool performance, hole quality, and productivity. This study aimed to investigate the drilling of SAE 1020 steel using a High-Speed Steel (HSS) drill under varying feed rates (f) (0.06, 0.09, 0.12, and 0.15 mm/rev). A constant spindle speed of 2000 rpm was used for all experiments, with a 5 mm diameter drill, and a chip-breaking drilling cycle was employed. The stopping criterion for the experiments was either the completion of 120 drilled holes or tool failure. The following output variables were analyzed: cutting tool wear, RMS electrical current required by the machine, and chip morphology. For this, flank wear (VBBmax) was monitored using a digital microscope, a low-cost RMS electric current acquisition system based on a Hall effect sensor was employed to capture real-time data, and the chips generated during the process were analyzed using scanning electron microscopy (SEM). The results indicated that feed rate significantly affected tool wear, with the highest feed rate condition leading to sudden drill failure after only 24 holes. Additionally, the average electric current values showed an increasing trend with higher feed rates. Chip morphology analysis revealed that feed rate not only altered the shape of the chips but also highlighted the presence of built-up edge formation during the drilling process. In conclusion, the feed rate had a substantial impact on all output variables, including tool wear, electric current, and chip formation. Based on the findings, lower feed rates are recommended when drilling with a 5 mm diameter drill under similar conditions to those used in this study.

Keywords Drilling · Wear · RMS electric current · Chip morphology

L. L. da Silva Araújo · P. A. de Oliveira · M. G. C. B. Barbosa (✉)
Department of Mechanical Engineering, Campus Ministro Petrônio Portella, Federal University of Piauí (UFPI), Teresina, Piauí, Brazil
e-mail: marcosguilherme@ufpi.edu.br

P. A. de Oliveira · M. V. da Silva Mendes · R. M. Monção · M. G. C. B. Barbosa
Postgraduate Program in Materials Science and Engineering, Campus Ministro Petrônio Portella, Federal University of Piauí (UFPI), Teresina, Piauí, Brazil

M. C. Gomes et al. (eds.), *Innovations in Brazilian Machining*, Lecture Notes in Mechanical Engineering, https://doi.org/10.1007/978-3-031-92819-2_6

1 Introduction

Drilling is a widely used machining operation in the industry [1]. It is a process of considerable economic importance, as it is mainly present in the final stages of manufacturing mechanical components [2]. It is a complex process, as the chip is formed inside the hole, and its removal can often be challenging. In drilling operations that do not use pre-drilling, the process can become even more difficult. According to Kumar et al. [2], the complexity of the drilling process is due to the drill geometry and plastic deformation of the material, with the cutting speed (v_c) being almost zero at the center of the drill. Additionally, the speed and rake angle vary as they move away from the drill's center, with the cutting edge having a higher peripheral cutting speed (v_c). According to Rivero et al. [3], the complexity of drilling is also associated with the drill's center, where cutting is combined with extrusion.

Given the level of mechanical stress that this type of tool endures, the material used to manufacture the drills for specific situations is also a crucial factor. High-speed steel (HSS) remains one of the materials widely used in twist drills. The main advantage of high-speed steel tools is their good vibration resistance, making them suitable for the drilling process. According to Teixeira [4], it can avoid mechanical shocks during drilling and handle different lubrication conditions. These characteristics can mitigate tool wear or failure.

Moreover, evaluating the progression of this wear under certain cutting conditions is extremely important to assess tool life, which is directly related to production costs and hole quality. For this, according to Barbosa et al. [5], a tool life analysis is carried out, and an end-of-life criterion can be established, such as maximum flank wear (VB_{Bmax}) or cutting tool fracture. Additionally, other output parameters can be evaluated, such as the machine tool's electric current and chip morphology.

Regarding wear types, they can occur in the form of craters, flanks, and even fractures. Flank wear is characterized by the gradual loss of material from the clearance surface due to the friction between the cutting tool and the machined surface [5]. Crater wear, according to Childs et al. [6], significantly reduces tool life due to plastic deformation or fracture. The cutting edge fracture occurs due to large machining forces [7].

The efficiency and durability of the cutting tool are also related to the workpiece material, as cutting parameters can vary for different materials. Situations that require low cutting speed (v_c), for example, can generate built-up edge (BUE) on the drill, leaving the hole in poor condition [8]. However, BUE formation can occur even at higher v_c for some materials, resulting in increased cutting forces and reduced tool life [9, 10]. Therefore, it is extremely important to individually evaluate the machining conditions for a specific cutting tool and a particular material to be machined, as different behaviors may occur in each situation.

In this context, this work aimed to evaluate the drilling process of SAE 1020 steel under different feed conditions and a fixed cutting speed (v_c) using high-speed steel drills, with the output variables being flank wear of the cutting tool, the RMS current required by the machine tool, and chip morphology.

2 Materials and Methods

The drilling tests were conducted on a vertical Machining Center, model ROMI D600, which features a maximum spindle rotation of 8000 rpm and a main motor power of 14.71 kW. High-speed steel (HSS) drills (PointTeq—BOSCH) with a diameter of 5 mm were used for the tests. These tools are manufactured in accordance with DIN 338 [11], classified as type N, and have a tip angle of 135°. The machined material was SAE 1020 steel in a standardized condition. Figure 1 illustrates the geometric characteristics of the workpiece, cutting tool, and the procedure for wear image acquisition.

The experiments were conducted with varying feed rates, while all other input parameters were held constant. A drilling cycle with chip breaking was employed (G73 function—FANUC), using a retraction value of 3 mm. Table 1 details the cutting conditions used, with the cutting speed (v_c) determined based on the tool manufacturer's recommendations and the feed rates established through preliminary tests.

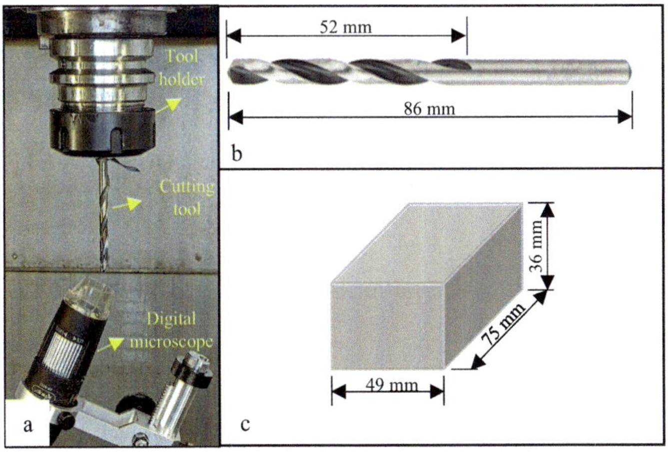

Fig. 1 Experimental setup: **a** digital microscope fixed to the machine tool table to monitor wear; **b** details of the cutting tool; **c** dimensions of the workpiece

Table 1 Cutting conditions

Cutting condition	f (mm/rev)	v_c (m/min)	Drilling depth (mm)
a	0.06	31.42	17
b	0.09		
c	0.12		
d	0.15		

Throughout all tests, the wear on the cutting tool and the RMS electrical current required by the machine tool were continuously monitored. Additionally, chips generated under the different cutting conditions were collected for morphological analysis. The machining of 120 holes was defined as the stopping criterion for the tests.

To monitor the wear behavior of the cutting tools, a digital microscope was mounted on the machine tool table and positioned to capture images of the drills' clearance surfaces (Fig. 1). Image capture was performed without removing the tool from the machine. The machine was programmed to position the tool in front of the microscope for image acquisition. This process was repeated until the tests reached the stopping criterion or the tool failed. The images were used to measure maximum flank wear (VB_{Bmax}) and to perform microscopic analysis of the worn surfaces.

The RMS electrical current required by the machine tool during the tests was monitored using a low-cost setup that included a non-invasive electrical current sensor (Hall Effect sensor), model SCT-013-000, with a range of 0–100 A; a data acquisition board with resistors for voltage division, a load resistor, and a filter capacitor; and an Atmega 2560 microcontroller. The current sensor was installed on one of the main power phases of the machine tool. Figure 2 shows the measured RMS electrical current signal, where the equipment's sensitivity to the machine's axis movements is evident. The signal peaks correspond to the rapid axis movements (G0 function—FANUC), while the highlighted regions in Fig. 2 correspond to movement at the programmed feed rate (G1 function—FANUC), where material removal occurs.

For the analysis of electrical current data, only the data corresponding to the region where material removal occurred were considered. A Quanta 450 FEG—FEI scanning electron microscope was used to capture images of the chips and evaluate their morphology.

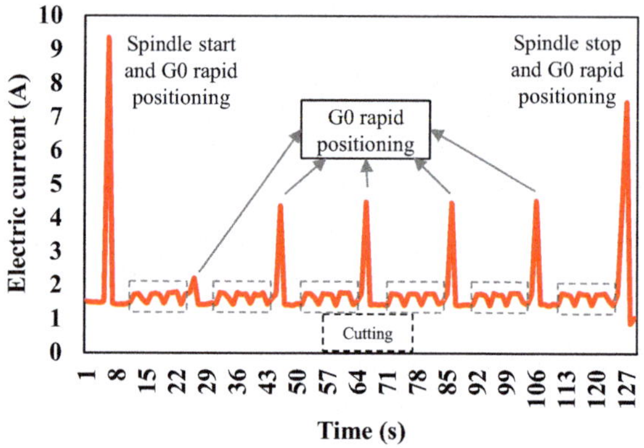

Fig. 2 RMS Electrical current signal monitored during 6 holes machining for cutting condition (a)

3 Results and Discussion

3.1 Wear

The maximum flank wear (VB_{Bmax}) of the cutting tools was measured at the conclusion of the tests. For cutting conditions (a) and (b), this analysis was performed after machining 120 holes. However, for cutting conditions (c) and (d), wear measurements were conducted after machining 108 and 24 holes, respectively, due to tool breakage, which prevented the tools from reaching the experiment's predetermined stopping criterion.

Figure 3 presents the average measured VB_{Bmax} values. It is evident that increasing the feed rates led to greater tool wear, becoming a critical factor for tool failure when feed values were set to 0.12 mm/rev and 0.15 mm/rev. In these cases, the tools broke before reaching the stopping criterion.

Figure 4 shows the condition of the cutting tools' clearance surfaces after the experiments. Under cutting conditions (a) and (b), flank wear and material buildup near the cutting edge were observed after machining 120 holes. For conditions (c) and (d), in addition to material accumulation, damage at the tool nose was evident, likely contributing to tool breakage.

These findings demonstrate the impact of feed rate on increased wear and tool failure. Higher feed values result in increased axial forces, which are directly linked to plastic deformation at the bottom of the hole [1, 7]. Consequently, greater material adhesion occurs near the cutting edge, and material accumulation on the rake surface increases, leading to a higher torsional moment [12, 13]. Thus, it can be concluded that the elevated mechanical demands caused by higher feed rates are responsible for the greater process of severity.

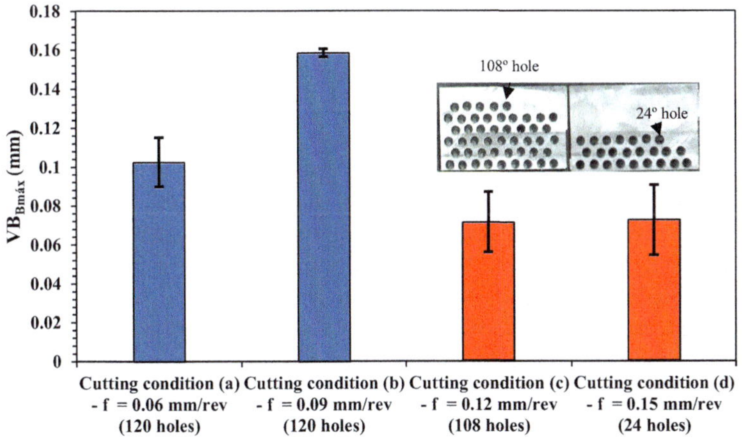

Fig. 3 Maximum flank wear for different cutting conditions

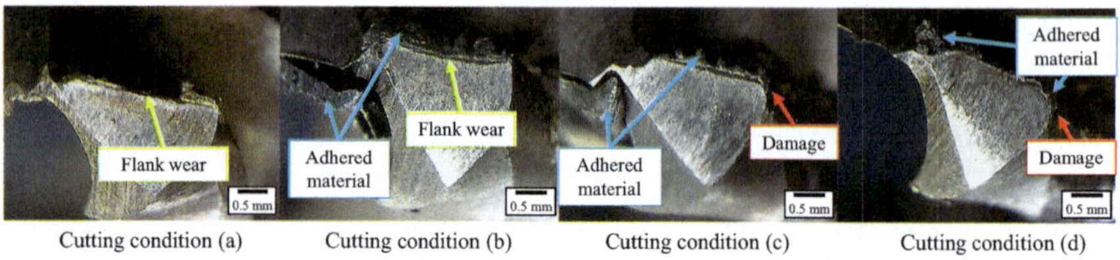

Fig. 4 Clearance surface of cutting tools

It is also important to emphasize that the experiments employed a drilling cycle with chip breaking to protect the cutting tool from material accumulation effects, particularly in the chip exit flutes. Nonetheless, material accumulation and adhesion were still significant, especially at higher feed rates. Therefore, in cases where drills have relatively small diameters (5 mm), adopting lower feed rates is advisable to preserve the cutting tool's integrity.

3.2 *Electric Current*

The RMS electrical current was monitored during the tests using a non-invasive sensor attached to one of the main power cables of the machine tool. Both active and passive movements performed during the machining of the holes were recorded, as shown in Fig. 2.

For the analysis of the results, only data corresponding to the drilling cycle were considered. The arithmetic means of these values, along with their dispersions, are presented in Fig. 5. The low-cost data acquisition system used was sensitive to variations in feed rates. The trend observed was an increase in the electrical current required by the machine tool as the feed rate increased. This supports the notion that machining efforts increase with higher feed rates, as the electrical current required by the machine tool is proportional to the machining efforts [5, 14, 15].

Additionally, Fig. 5 shows that the standard deviation of the measured values increased with higher feed rates. This finding is noteworthy and is likely linked to the characteristics of the monitoring system. The system's signal acquisition rate was 1 Hz, meaning one measurement was taken every second. As a result, the system may not have captured some data points during periods of higher feed rates, leading to a greater standard deviation. This effect was further amplified by the drilling cycle adopted, as the data acquisition system was unable to record all movements between the feed steps that removed material, and the retraction steps used to break chips at the highest feed rates tested.

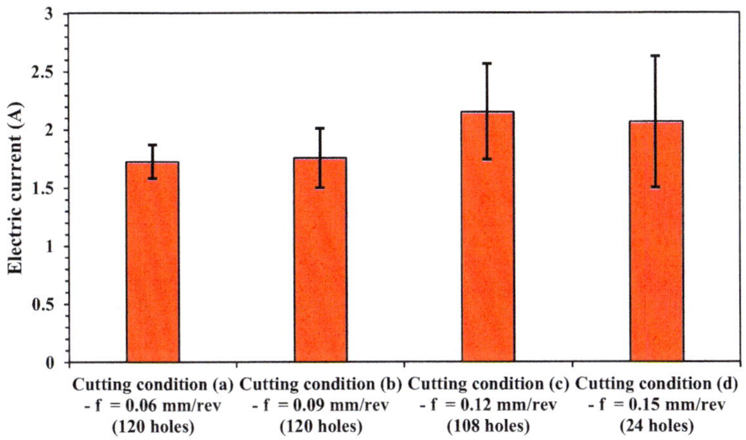

Fig. 5 Average RMS electrical current values

3.3 *Chip Morphology*

Chip morphology plays a crucial role in understanding the interactions at the chip-tool interface, providing valuable insights into cutting tool wear and other machining outcomes. In Fig. 6, macrographs of the chips formed under various cutting conditions are shown.

For cutting condition (a), where the lowest feed rate was used, the chip formed was thinner and longer compared to the chips from other cutting conditions? This behavior aligns with expectations [12], as the drilling strategy included chip breaking, and the feed rate was increased in the other cutting conditions relative to condition (a). The variation in chip shape played a significant role in the cutting tool's wear behavior. Thicker chips exhibited a bluish hue, indicating higher cutting temperatures. Additionally, material adhesion near the cutting edge was more pronounced at higher feed rates (Fig. 4), which contributed substantially to increased tool wear and eventual tool failure.

Continuous chips were formed during machining with the two lowest feed rates, as shown in Fig. 7, which was expected due to the machining of low-carbon steel. At the two highest feed rates, evidence of lamellar chip formation appeared on the chip

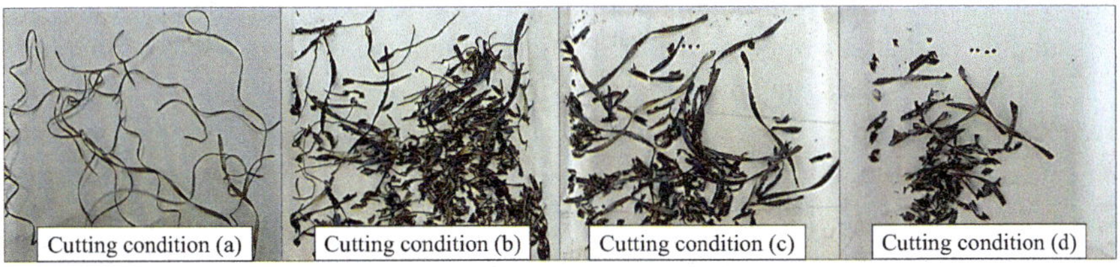

Fig. 6 Chips formed during tests under the different cutting conditions adopted

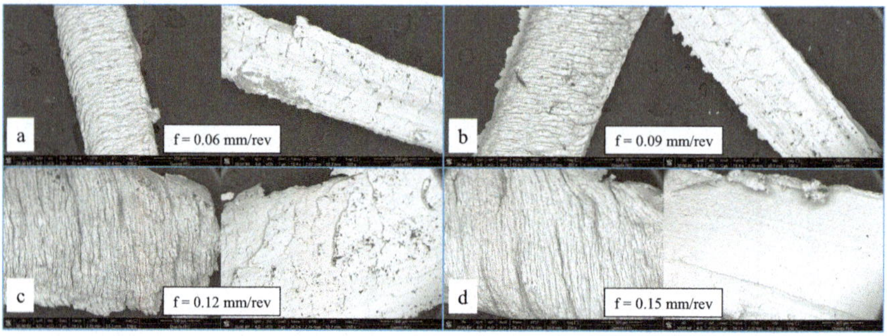

Fig. 7 Chip morphology under different cutting conditions: **a** cutting condition (a); **b** cutting condition (b); **c** cutting condition (c); **d** cutting condition (d)

front surface, likely induced by the greater deformation imposed by these cutting conditions. Additionally, images of the chip's posterior surface revealed material adhered to areas near the cutting edge, with alternating rough and smooth zones. According to Trent and Wright [16], this phenomenon occurs due to fractures along and across flow lines when cutting with a built-up edge (BUE).

These observations align with the findings on tool wear, with material adhesion identified as the predominant cause of tool wear and failure across the different cutting conditions used in the experiments.

4 Conclusions

Through the results presented and discussed, it is possible to conclude that:

i. The increase in feed rates caused greater wear and cutting tools failure, with the accumulation of material in the regions close to the cutting edge being the probable cause of this behavior.
ii. The measured RMS electric current values demonstrate that machining efforts increased with the adoption of higher feed rates. This, in conjunction with material adhesion in the drill flutes, likely led to an increase in axial force and torsional moment, ultimately resulting in cutting tool breakage under conditions with elevated feed rates.
iii. The electrical current acquisition system, due to its low acquisition rate, was unable to detect some signals when higher feed values were used, contributing to the highest calculated standard deviation.
iv. Regarding chip formation, continuous chips were observed under the two lower feed conditions, whereas a tendency for lamellar chip formation emerged at higher feed rates. This behavior can be attributed to the greater deformation

induced by the increased feed values. Additionally, analysis of the chip's posterior surface confirmed the presence of a built-up edge (BUE) during the tests, corroborating the influence of material adhesion on tool wear and failure.

References

1. Sujan Kumar M, Deivanathan R (2021) Effect of process parameters on drilling—an overview. Mater Today Proc 46(2):1401–1406. https://doi.org/10.1016/j.matpr.2021.02.574
2. Suresh Kumar B, Baskar N, Rajaguru K (2020) Drilling operation: a review. Mater Today Proc 21(1):926–933. https://doi.org/10.1016/j.matpr.2019.08.160
3. Rivero A, Aramendi G, Herranz S, López de Lacalle LN (2006) An experimental investigation of the effect of coatings and cutting parameters on the dry drilling performance of aluminium alloys. Int J Adv Manuf Technol 28(1–2):1–11 (2006). https://doi.org/10.1007/s00170-004-2349-3
4. Teixeira CR (1995) Influência dos erros geométricos de afiação nas brocas helicoidais de aço rápido na qualidade dos furos produzidos (in portuguese). Federal University of Santa Catarina. Technology Center. Postgraduate Program in Mechanical Engineering
5. Barbosa MGCB, Hassui A, de Oliveira PA (2021) Effect of cutting parameters and cutting edge preparation on milling of VP20TS steel. Int J Adv Manuf Technol 116(9–10):2929–2942. https://doi.org/10.1007/s00170-021-07654-7
6. Childs T, Maekawa K, Obikawa T, Yamane Y (2000) Metal machining, 1st edn. Elsevier. https://doi.org/10.1016/C2009-0-23990-0
7. Ferraresi D (1970) Fundamentos da Usinagem dos Metais, 1st edn. Blucher, São Paulo
8. Vilanova AM (2023) Investigação sobre a qualidade dos furos gerados na furação dos aços inoxidáveis austenítico Core® 304L e super austenítico Ultra® 904L (in portuguese), Master's dissertation. Federal University of Rio Grande do Sul. School of Engineering. Postgraduate Program in Mechanical Engineering
9. Klocke F (2011) Manufacturing processes 1, 1st edn. Springer, Berlin. https://doi.org/10.1007/978-3-642-11979-8
10. Nomani J, Pramanik A, Hilditch T, Littlefair G (2013) Machinability study of first generation duplex (2205), second generation duplex (2507) and austenite stainless steel during drilling process. Wear 304(1–2):20–28. https://doi.org/10.1016/j.wear.2013.04.008
11. Deutsches Institut für Normung (1978) DIN 338—Parallel shank twist drills, jobber series. Berlin
12. Diniz AE, Marcondes FC, Coppini NL (2013) Tecnologia da Usinagem dos Materiais, 8th edn. Artliber, São Paulo
13. Ertunc HM, Oysu C (2004) Drill wear monitoring using cutting force signals. Mechatronics 14(5):533–548. https://doi.org/10.1016/j.mechatronics.2003.10.005
14. de Oliveira PA, de Menezes Pereira LMP, Monção RM, de Paiva RL, Barbosa MGCB (2024) Tool wear, surface roughness, electric current, and chip morphology in the turning of AISI 1045 steel with minimum quantity lubrication (MQL) technique. Int J Adv Manuf Technol 133(11–12):5743–5759. https://doi.org/10.1007/s00170-024-14094-6
15. Machado AR, Abrão AM, Coelho RT, da Silva MB (2015) Teoria da Usinagem dos Materiais, 3rd edn. Blucher, São Paulo
16. Trent E, Wright P (2000) Metal cutting, 4th edn. Butterworth-Heinemann, Oxford

Investigation of the Impact of Different Trajectory Angles on the Milling of ER70S-6 Parts Obtained by WAAM-CMT

Victor Lacort Natividade, Maksym Ziberov, Álisson Rocha Machado, and Déborah de Oliveira

Abstract Additive manufacturing (AM) has become increasingly relevant in global manufacturing, offering a cost-effective and straightforward way to produce complex parts that are difficult to achieve using conventional methods. However, issues such as dimensional accuracy and surface quality of AM-manufactured parts often require post-processing, highlighting the importance of machining in the production line. In the case of Wire and Arc Additive Manufacturing (WAAM), these challenges are even more pronounced, once the method uses high thermal inputs to create metal parts with overlapping layers, which can lead to anisotropy and discontinuities. Therefore, considering the anisotropic characteristics of AM metals, this study investigates how the cutting path angle influences the surface quality after end milling of parts manufactured with ER70S-6 wire. For this purpose, a carbide milling cutter with titanium aluminum nitride (TiAlN) coating, with a diameter of 5 mm and four cutting edges, and five different cutting angles, ranging from 0° to 90° in 15° increments, was used. The analyses revealed that the slots milled at 45° presented the best surface quality and lowest average roughness value Rq = 0.22 μm, while the slots at 15° had the worst qualitative and quantitative results, Rq = 0.53 μm. It was concluded that cutting angles greater than 45° produce better quality machined surfaces compared to lower angles, becoming the best recommendation for milling parts manufactured by WAAM.

Keywords WAAM · Milling · Cutting parameters · Milling direction · Surface finishing

V. L. Natividade · M. Ziberov · D. de Oliveira (✉)
Department of Mechanical Engineering, University of Brasilia, Brasilia, Brazil
e-mail: oliveira.deborah@unb.br

Á. R. Machado
Mechanical Engineering Graduate Program, Pontifical Catholic University, Curitiba, Brazil

M. C. Gomes et al. (eds.), *Innovations in Brazilian Machining*, Lecture Notes in Mechanical Engineering, https://doi.org/10.1007/978-3-031-92819-2_7

1 Introduction

Product manufacturing in industry traditionally involves the molding of materials through processes such as melting, forming, removing, and assembling, using technologies such as casting, injection molding, machining, and forging. However, since the 1980s, additive manufacturing (AM) has emerged as a new approach, based on the addition of material to form parts with complex geometries [1]. This revolutionary method has influenced both academia and industry, facilitating manufacturing processes and simplifying design steps such as prototyping, in addition to enabling the production of parts that would be unfeasible using conventional methods. As an additive process, AM also minimizes material waste, making it a more economical option [2].

Within AM technologies, the Wire and Arc Additive Manufacturing (WAAM) stands out for using electrode wire as an additive material. The Cold Metal Transfer (CMT) method, an evolution of Gas Metal Arc Welding (GMAW), is widely used to manufacture complex parts, employing welding robots to superimpose layers of material [3]. The importance of AM involving metals is reinforced by its favorable mechanical properties, which are widely applicable in engineering, offering new metal manufacturing solutions and alternatives to conventional methods [4].

This type of part was studied by the authors [5], focusing on the microstructure of the part obtained by AM, they noticed a microstructural composition similar to cast parts, with the presence of dendrites, characteristic of materials from welding, but with a uniform orientation, originating from the direction of deposition of the material. In this sense, to have less influence of the thermal gradients it is possible to deposit the material with temperature control [6] or improve the mechanical and geometrical characteristics by varying the deposition parameters [7] or the deposition paths according to the part geometry [8].

The ER70S-6 welding wire is commonly used in the welding universe due to its low cost and effective performance in joining low-carbon steel parts, allowing a wide range of applications [9, 10]. However, to solidify its position in the global industry, it is crucial to deepen knowledge about the characteristics of the parts produced, standardizing manufacturing, inspection and testing parameters to ensure quality, finish and adequate geometric tolerances [11].

To achieve the project requirements post-processing became necessary [12] and different methods can be applied for improving the finish and surface quality of complex parts [13]. Aiming to improve the surface of WAAM parts, the authors [14–16] milled ER70S-6 WAAM-CMT parts, each one to evaluate the influence of a different cutting parameter on the machined surface. On their work they applied a similar experimental apparatus that will be used in this work. A carbide mill, TiAlN coated, with 5 mm in diameter and 4 cutting edges, cutting fluid BIO10E and a CNC vertical milling machine.

The authors [14, 15] analyzed the influence of different cutting fluid flow rates, as well as different nozzle positioning arrangements. For [14], the best roughness result Rq was 0.61 μm, for a flow rate of 10 l/min and a cutting fluid nozzle positioned in the direction of tool feed. The author [15] achieved a slightly lower value, Rq = 0.58, for two cutting fluid nozzles positioned at 45° in relation to the direction of tool feed and a compressed air nozzle also at 45°. Worth mentioning that, for these studies, the authors used a cutting speed vc = 10 m/min and a cutting direction perpendicular (90°) to the direction of material deposition. The author [16], made some changes to the parameters, varied the cutting speed to vc = 10 m/min and vc = 40 m/min, used the nozzle and flow arrangement adopted by [15], and varied the cutting direction of the tool at 0° and 90°, regarding the deposition direction. In this work the improvement in the roughness result was significant, reaching Rq = 0.29 μm for the best slot, referring to the cut made parallel (0°) to the material deposition direction and higher cutting speed vc = 40 m/min.

Therefore, it is possible to notice that improvements are being achieved by studying the milling of WAAM parts, and that the direction of the deposition might have great influence on the final finishing. Besides this topic still little explored, it offers opportunities for significant advances in metal AM, both in academic and industrial environments, promoting more reliable and economically viable results. In this sense, the main objective of this work is to evaluate the impact of different cutting tool path angles on the surface finish of ER70S-6 parts, focusing on the measurement of root mean square roughness (Rq) and qualitative surface analysis, by evaluating images of the surface.

2 Materials and Methods

The study is based on the machining process, performed by end milling an ER70S-6 part obtained by WAAM-CMT, details on the deposition method can be found in the work of [9]. This material presents a low percentage of alloying elements added, with composition in weight being 0.07% C, 0.85% Si and 1.50% Mn.

Once the objective of the work was to evaluate the influence of the machining path in respect of the layer deposition direction, the cutting trajectory angle was varied between 15° and 75° in steps of 15° in relation to the material deposition direction. The part has dimensions equal to 5 mm thick, 40 mm wide and 160 mm long. Figure 1 presents the schematic representation.

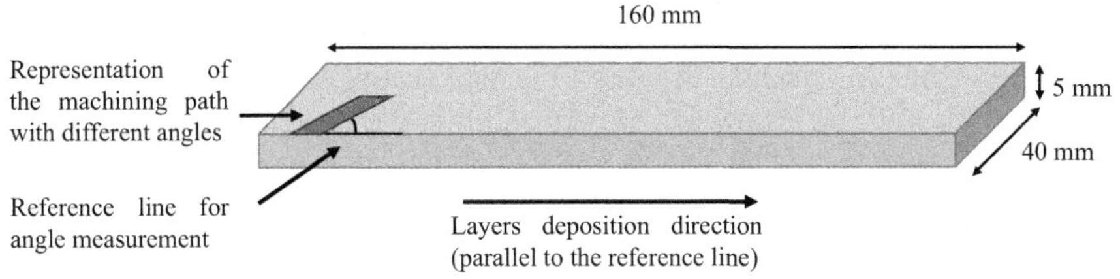

Fig. 1 Schematic view of the machined part and trajectory

The machine tool used is a vertical machining machine of the VEGA XH7132 model. A single tool manufactured all the slots, a carbide milling cutter coated with Titanium Aluminum Nitride (TiAlN), model G9A69050 from the manufacturer YG-1 with 5 mm diameter and 4 cutting edges. The cutting fluid used was the BIO10E model from the manufacturer Biolub Química, through two cutting fluid nozzles positioned at 45° in relation to the tool feed direction, as well as a nozzle for injecting compressed air.

The machining parameters adopted, based on the studies by [14–16] are the cutting speed vc = 40 m/min, machining depth ap = 0.25 mm, feed per tooth fz = 0.020 mm and rotation n = 2548 rpm. It is worth mentioning that the G code applied to the CNC machine tool was developed with the help of Autodesk's Fusion 360 software, through which the tool's input and output coordinates were identified for each slot and then inserted into the G code.

With output variables corresponding to the roughness Rq, measured using a portable roughness meter of the SJ-201 model from the manufacturer Mitutoyo, with a cut-off of 0.8 mm and a sampling length of 4.00 mm, three measurements were performed, and the results will be presented by the mean values with two times the standard deviation. The images for the qualitative analysis were obtained from a 5x magnification lens, using a microscope of the LEXT OLS4100 model.

3 Results

For the roughness measurements of each slot, 4 measurements of the Rq parameter were performed, and their averages and respective standard deviations were calculated. Table 1 presents the values obtained, referring to the replica performed at 45° and the test performed at 15°, respectively. It also presents the results measured for the surface of the part prior to the machining with the selected parameters.

Figure 2 illustrates graphically the data in Table 1, the behavior of roughness according to the change in the machining path angle. For angles greater than or equal to 45°, the values have a similar behavior, present in the range between 0.20 and 0.35 μm. Furthermore, when the angle is less than 45°, the roughness tends to increase, as can also be seen, reaching values close to 0.60 μm when analyzing the standard deviation, approximately twice the lowest result measured for the 45° path. This shows that the impact of the change in the machining angle in relation to the material deposition direction is large, justifying the need to study all the cutting parameters that, consequently, determine the final quality of the entire process.

Thus, based on studies previously carried out in the same line of research, it is possible to observe the improvement in the final roughness results Rq presented in Fig. 3, which illustrates a bar graph with the lowest values obtained for each study in chronological order. The current work resulted in a roughness approximately three times lower than the study by [14] and lower than all the others. Indicating that the change in the trajectory angle, as well as the adoption of adequate parameters, result in better design quality.

Table 1 Rq values for the slots and part surface

Slot		Rq_1	Rq_2	Rq_3	Rq_4	Mean	Standard deviation
75°	Trial	0.27	0.26	0.36	0.28	0.29	0.05
	Replica	0.23	0.28	0.27	0.28	0.27	0.02
60°	Trial	0.28	0.23	0.23	0.24	0.25	0.02
	Replica	0.33	0.34	0.27	0.34	0.32	0.03
45°	Trial	0.28	0.28	0.25	0.24	0.26	0.02
	Replica	0.24	0.22	0.21	0.20	0.22	0.02
30°	Trial	0.33	0.26	0.33	0.31	0.31	0.03
	Replica	0.57	0.39	0.54	0.55	0.51	0.08
15°	Trial	0.54	0.43	0.56	0.58	0.53	0.07
	Replica	0.49	0.43	0.45	0.50	0.47	0.03
Surface	Trial	0.67	0.68	0.94	0.96	0.81	0.16
	Replica	1.94	1.77	1.11	0.95	1.44	0.49

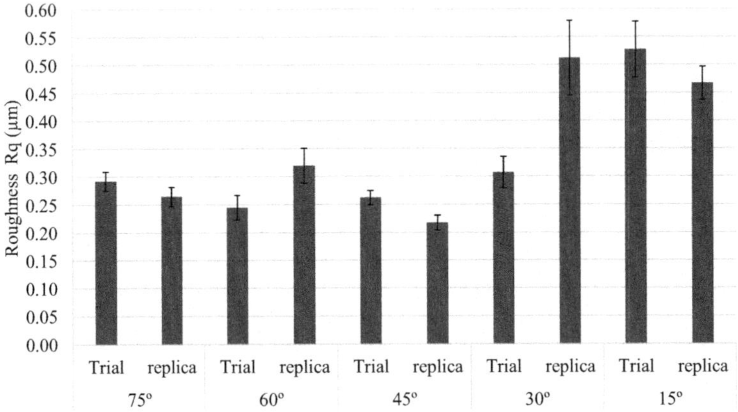

Fig. 2 Average Rq values calculated for the slots

For surface analysis, the images obtained through a microscope with a 5x magnification lens of all slots are shown in Fig. 4, with a green border highlighting the slot with the lowest roughness value Rq and the slot with the highest value in red. It can also be seen that the surface quality of the slots follows the same logic presented in Fig. 2, in which angles below 45° result in lower quality surfaces, with clear and inhomogeneous feed marks, the opposite of what is observed for the other slots, with more homogeneous and visually clean surfaces.

Figure 5 aids the analysis by comparing the best and worst slots, on can note that the surface of the replica performed at 45° has a more uniform behavior when compared to the test performed at 15°. The feed marks follow a pattern for the 45° angle, with less intense variations, indicating a uniform surface in relation to the

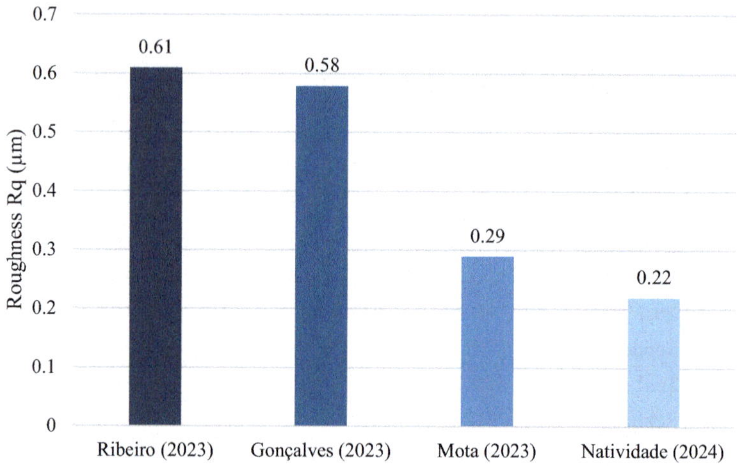

Fig. 3 Comparison with other work on milling ER70S-6 obtained by WAAM

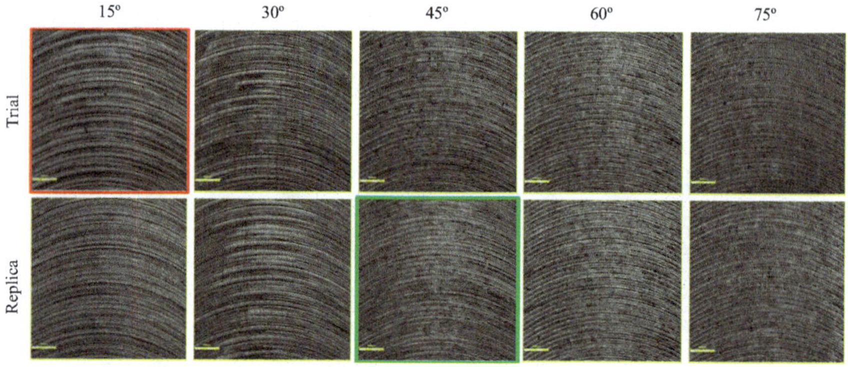

Fig. 4 Images of slots surfaces after machining

peaks and valleys generated with each rotation of the tool. The behavior is opposite when focused on the 15° slot, with clear and non-uniform feed marks, varying the behavior throughout the analyzed area. Regions that indicate variations between peaks and valleys are more intense. As well as regions that indicate material pullout and points that indicate pores in the structure, probably resulting from the manufacturing method. The identification of those defects is important once the porosities will interfere with the fatigue resistance [17] and mechanical behavior [18].

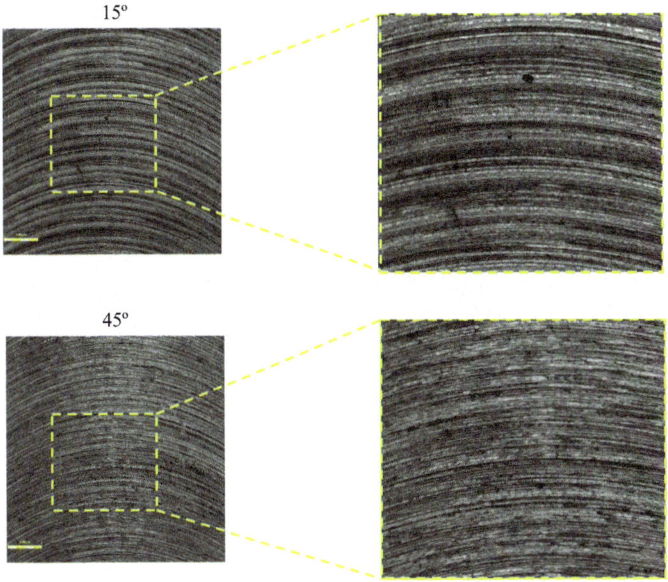

Fig. 5 Detail of the surface condition for 15° and 45°

4 Conclusions

From the experimental results, it is concluded that the best path angle to be adopted for end milling of ER70S-6 parts manufactured by WAAM-CMT is 45° in relation to the material deposition direction, with Rq = 0.22 μm. This presents an improvement of 59% in relation to the worst slot (15°), Rq = 0.53 μm. In addition, adopting angles between 90° and 45° also corresponds to a good design choice, since they presented similar behavior to the 45° slot, with good surface quality, homogeneous behavior and subtle feed marks, without noticeable damage. The opposite is observed for angles between 0° and 45°, and this range should be avoided, since they have the worst results, both qualitatively and quantitatively.

Acknowledgements The authors are grateful to the University of Brasilia and CNPq, grant numbers, 402730/2023-2, 443148/2023-6 and 200018/2025-6 for the financial support, and LAETA/INEGI—Associate Laboratory of Energy, Transports and Aerospace/Institute of Science and Innovation in Mechanical Engineering and Industrial Engineering.

References

1. Volpato N (2017) Manufatura aditiva. Tecnologias e aplicações da impressão 3D. Blücher, São Paulo

2. Wiltgen F (2019) Protótipos e prototipagem rápida aditiva sua importância no auxílio do desenvolvimento científico e tecnológico. In: 10th Brazilian manufacturing engineering congress. São Carlos, pp 1–6
3. Laukik PR, Ravindra VT (2021) Wire arc additive manufacturing: a comprehensive review and research directions. J Mater Eng Perform 30(7):4768–4791
4. Dilberoglu UM, Gharehpapagh B, Yaman U, Dolen M (2017) The role of additive manufacturing in the era of industry 4.0. Procedia Manuf 11:545–554
5. Laue R, Colditz P, Möckel M, Awiszus B (2022) Study on the milling of additive manufactured components. Metals 12(7):1167
6. De Araújo Soares MA, Novelino ALB, Ziberov M (2024) Geometry study on 410NiMo alloy parts printed by WAAM-CMT. In: de Oliveira D, Ziberov M, Rocha Machado A (eds) ABCM series on mechanical sciences and engineering. COBEF 2023. LNCS. Springer, Cham, pp 114–125
7. Vieira RDAM, Liskevych O, Ziberov M (2024) Influence of deposition strategy in CMT-based wire and arc additive manufacturing on part geometry: a literature review. Soldagem Inspeção 29(e2910)
8. Vieira RDAM, Liskevych O, De Oliveira D, Ziberov M (2024) Influence of parameter variation and interlayer temperature control in wall angle, curvature and measurement methodology of ER70S-6 parts obtained by WAAM. Manuf Lett 42:40–45
9. Novelino ALB, Carvalho GC, Ziberov M (2022) Influence of WAAM-CMT deposition parameters on wall geometry. Adv Ind Manuf Eng 5:100105
10. Zhai W, Guo Y, Aishwarya, Canturri C, Shandro R, Zhou W (2024) Wire arc additive manufacturing of ER70S-6/S355 bimetal component. Mater Sci Eng A 900:146498
11. Gomes JFB, Wiltgen F (2020) Avanços na manufatura aditiva em metais: técnicas, materiais e máquinas. Revista Tecnologia 41(1):1–16
12. Gomes MC, Dos Santos AG, De Oliveira D, Figueiredo GV, Ribeiro KSB, De Los Rios GAB, Da Silva MB, Coelho RT, Hung WNP (2022) Micro-machining of additively manufactured metals: a review. Int J Adv Manuf Technol 118:2059–2078
13. De Oliveira D, Gomes MC, Dos Santos AG, Ribeiro KSB, Vasques IJ, Coelho RT, Da Silva M, Hung WNP (2023) Abrasive and non-conventional post-processing techniques to improve surface finish of additively manufactured metals: a review. Prog Addit Manuf 8:223–240
14. Ribeiro GM (2023) Verificação da influência do posicionamento do bocal do fluido de corte no fresamento de peças de ER70S-6 obtidas por manufatura aditiva via CMT, Monograph. Department of Mechanical Engineering, University of Brasilia, Brasilia, pp 1–73
15. Gonçalves MV (2023) Fresamento de peças de ER70S-6 obtidas por manufatura aditiva com diferentes atmosferas lubri-refrigerantes, Monograph. Department of Mechanical Engineering, University of Brasilia, Brasilia, pp 1–63
16. Mota EN (2023) Avaliação da qualidade superficial no fresamento de peças de ER70S-6 obtidas por MADA em função da velocidade de corte e da direção de deposição, Monograph. Department of Mechanical Engineering, University of Brasilia, Brasilia, pp 1–96
17. Araujo LC, Ferreira JLA, Ziberov M, Araujo JA (2023) Assessing fatigue in materials with small defects: a new multiaxial model based on principal stress amplitudes. Procedia Struct Integr 57:144–151
18. Silva CP, Silva GP, Santos MC, Ziberov M, Malcher L (2024) Hybrid search methodology for mechanical characterization of material produced via WAAM assuming Gurson porous material. J Braz Soc Mech Sci Eng 46(266)

Calculation of Measurement Uncertainty Associated with Ra Roughness of FC 250 Gray Cast Iron After Grinding

Amanda Souza Abrão, Ana Caroline Teixeira Messias, Bruno Souza Abrão, Alcione dos Reis, Mayara Fernanda Pereira, and Rosemar Batista da Silva

Abstract Gray cast iron is widely used in several applications that require combination between low roughness Ra values and narrow dimensional tolerances. Therefore, grinding can be an alternative process to be used at the end of the production chain for manufacturing products having such requirements. Furthermore, to determine metrological traceability and improve reliability of measurements of surface roughness values, that allows to take best decisions, the comparison of the results obtained with other reference values are indispensable. This can be performed through the calculation of measurement uncertainty. Into this context, calculation of the measurement uncertainty associated with the surface roughness Ra parameter of FC 250 Gy cast iron after the grinding process was performed. Different cutting conditions were used (cutting speed (Vc) = 31.16 m/s, workspeed (Vw) = 7.5 m/min, number of passes of grinding wheel on the material surface (1 and 2) and radial depth of cut (ae) = 15 µm and 30 µm) and silicon carbide grinding wheel and under a coolant environment. The roughness Ra was measured on the ground surface of the workpiece and the measurement uncertainty was calculated by the GUM method. The results showed that the reduction in the number of passes together with the increase in the radial depth of cut resulted in the highest roughness Ra value of 0.78 µm. The measurement uncertainty was calculated for the two conditions: two passes of 15 µm and one pass of 30 µm, resulting in Ra = (0.73 ± 0.08) µm and Ra = (0.78 ± 0.06) µm, respectively. The variability of the readings was the greatest contribution to the final measurement uncertainty associated with the obtained Ra (by about 87.06% and 71.69% for two passes of 15 µm and one pass of 30 µm, respectively). Also, Ra surface roughness increased with radial depth of cut.

Keywords Grinding · Gray cast iron · Metrology · Ra surface roughness · Measurement uncertainty

---

A. S. Abrão (✉) · A. C. T. Messias · B. S. Abrão · A. dos Reis · M. F. Pereira · R. B. da Silva
Federal University of Uberlândia, Uberlândia, Minas Gerais, Brazil
e-mail: amandaabrao55@gmail.com

M. C. Gomes et al. (eds.), *Innovations in Brazilian Machining*, Lecture Notes in Mechanical Engineering, https://doi.org/10.1007/978-3-031-92819-2_8

1 Introduction

Among the different types of cast iron used in industry, gray cast iron is the most used in several industries, comprising a wide range of applications that varies from engine blocks to decorative items, as well as automotive brake discs. The latter are subjected to the grinding process to achieve the roughness, dimensional and macrogeometric deviations necessary for coupling with brake pads, given the interchangeability of mechanical elements [1].

The grinding process involves the removal of material from the workpiece through the action of thousands of abrasive grits having non-defined geometry and randomly orientation. It is usually performed at the end of the production chain to obtain the combination of low roughness values (Ra $\leq$ 1.60 μm) and narrow dimensional tolerances (international tolerance ISO IT06 to IT03) [2]. Grinding performance depends on the correct selection of cutting parameters such as radial depth of cut, workspeed and cutting speed, among others. In addition, for these situations that require a good finish, machining processes with a tool with a defined geometry cannot be very effective, so the grinding process emerges as an alternative to meet certain project requirements [2].

Ra roughness parameter is the most widely used to assess the surface quality of a component, also known as the arithmetic mean deviation. However, most studies that evaluate roughness do not address the calculation of measurement uncertainty for this parameter. Calculation of measurement uncertainty is a non-negative parameter that presents the dispersion of the values given to a measurand, based on information contained in standards, certificates and other references in the literature [3] and its evaluation helps to compare the results obtained with other reference values. Furthermore, the expression of measurement uncertainty assists in the metrological traceability of measurement results, as well as to allow comparison between them and other results presented in the literature [4]. Measurement uncertainty is usually based on the Guide methodology for calculation of the expression of measurement uncertainty (GUM). Therefore, this work seeks to analyze the influence of radial depth of cut on the roughness parameter Ra of gray cast iron and to do the calculation of measurement of uncertainty associated with the Ra roughness.

2 Methodology

The grinding process was carried out on a semi-automatic peripheral surface grinding machine from the Mello manufacturer, P36 model, that is driven by 2.25-kW nominal power and constant spindle speed of 2400 rpm. A Silicon Carbide (SiC) grinding wheel—straight type—with 39C60KVK designation, with dimensions of 248 mm (external diameter) $\times$ 127 mm (internal diameter) $\times$ 25.4 mm (width), from the NORTON—Saint Gobain Abrasives manufacturer was used. The experimental grinding system is shown in Fig. 1. The workpiece material selected for the grinding

Fig. 1 Experimental set-up for grinding of gray cast iron

trials was the gray cast iron FC 250 supplied by Tupy S.A. with a hardness of 207 HB [5] and dimensions of 47 mm (long) × 17 mm (high) × 15 mm (wide). Its chemical properties are shown in Table 1.

Cutting speed (Vs) of 31.16 m/s, workspeed (Vw) of 7.5 m/min and two values of radial depth of cut (ae) of 15 μm and 30 μm were the input parameters employed in this work. The test stop criterion was determined by the amount of material removal of 9.45 mm^3. Since two different radial depth of cut values were used, two grinding passes were performed using ae = 15 μm and only one grinding pass using ae = 30 μm. Prior to each test, the dressing operation was performed using a single diamond tip dresser. The dressing parameters were the grinding wheel overlap ratio (Ud) of 3, dressing speed (Vd) of 240 mm/min and 2 passes of a dressing depth (ad) of 30 μm by the dresser on the grinding wheel. In Fig. 1 is presented the Experimental set-up for grinding of gray cast iron.

All the grinding tests were carried out with a semi-synthetic vegetable oil-based (Blaser Swisslube VASCO 7000) cutting fluid at a dilution ratio of 1:19, which was delivered at flow rate of 8 L/min by the conventional coolant technique. The calculation of the uncertainty associated with the measurement of roughness (Ra) was performed using a rugosimeter from Mitutoyo, model SJ-201, with a resolution of 0.01 μm and a probe tip radius of 2 μm. Five measurements were performed under room temperature at the measurement site of 20 ± 1 °C in the Laboratory of Metrology (LEMETRO) from School of Mechanical Engineering at Federal University of Uberlândia (FEMEC-UFU). In addition, the roughness of the parts was

Table 1 Chemical composition of gray cast iron FC 250 [5]

C	Si	Mn	P	S	Cr	Ti	Sn	Mo
3.4 to 3.6	1.67	0.42	0.026	0.076	0.087	0.01	0.086	0.284

recorded on a granite table and the calculations of measurement uncertainty were performed using the GUM method based on the Guide for Expressing Measurement Uncertainty methodology. To apply the GUM method is necessary to follow the steps [2]:

1. To identify the measurand (output variable);
2. To identify the input variables (variables that can affect the measurement result of the measurand);
3. To identify the mathematical model of the measurement;
4. To apply the Law of Propagation of Uncertainties in the mathematical model;
5. To calculate the standard uncertainty associated with the input variables;
6. To calculate the combined standard uncertainty of the output variable;
7. To calculate the expanded uncertainty related to the output variable;
8. To determine the measurement result;
9. To calculate the contributions of the input variables to the final uncertainty.

3 Results and Discussions

In Fig. 2 are presented measurements of the roughness values Ra as a function of the number of passes and the radial depth of cut after grinding the gray cast iron.

When comparing both parameters, it can be noticed that the use of a pass from grinding wheel with a radial depth of cut of 30 μm was responsible for an increase of approximately 6.4% in the Ra roughness (Fig. 2). Under this condition there is a larger contact area between grinding wheel and workpiece surface that means a greater number of abrasive particles in contact with the workpiece in comparison with the lowest radial depth of cut. As a result, there is an increase in the material removal rate, greater equivalent chip thickness that worsens the tribological conditions and deteriorates the finish. A direct relationship between roughness and radial depth of

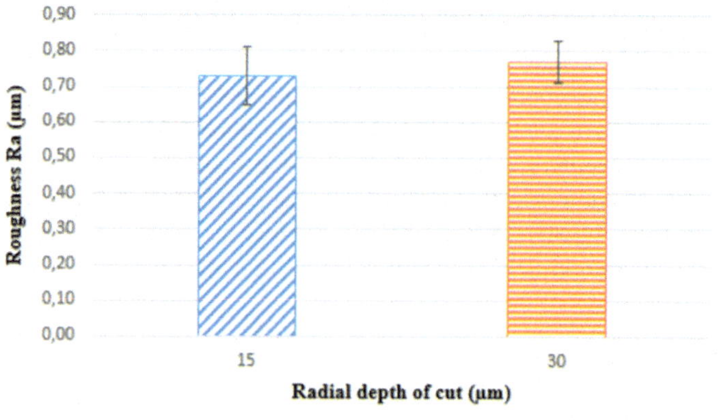

Fig. 2 Average values of the Ra parameter of gray cast iron FC 250 as a function of radial depth of cut

cut in the grinding of gray cast iron using a silicon carbide (SiC) grinding wheel was also observed by [6]. This author investigated two radial depth of cut values, 15 and 30 μm, and reported that roughness increased with radial depth of cut. According to Lima [6], grinding with low radial depth of cut with values similar to those used in this work leads results less number of abrasive grits removing material which consequently reduce heat generation in the grinding wheel-workpiece interface. This is favorable to provide low surface roughness, i.e. to improve surface finish.

In Tables 2 and 3 are presented the results on the calculation of the measurement uncertainty, for the grinding conditions of 2 passes of 15 μm and 1 pass of 30 μm, respectively.

Table 2 Ra uncertainty for 2 passes of 15 μm

Magnitude	Estimated	Probability distribution	Type of assessment	Degrees of freedom	Standard uncertainty	Contribution (%)
x	0.73000	t-student	A	1	0.04000	87.06441
ΔR	0.01000	Rectangular	B	∞	0.00289	0.45346
ΔC	0.01000	Normal	B	∞	0.00500	1.36038
ΔRp	0.01460	Rectangular	B	∞	0.00843	3.86639
ΔD	0.02000	Rectangular	B	∞	0.01155	7.25537
Combined standard uncertainty (uc)						0.04287
Effective degrees of freedom (Veff)						66,982.24462
Coverage factor (k)						1.96000
Expanded uncertainty (U)						0.08402

Ra = (0.73 ± 0.08) μm, for a coverage factor k of 1.96 and a confidence of 95%

Table 3 Ra uncertainty for 1 pass of 30 μm

Magnitude	Estimated	Probability distribution	Type of assessment	Degrees of freedom	Standard uncertainty	Contribution (%)
x	0.77500	t-student	A	1	0.02500	71.69487
ΔR	0.01000	Rectangular	B	∞	0.00289	0.95593
ΔC	0.01000	Normal	B	∞	0.00500	2.86779
ΔRp	0.01550	Rectangular	B	∞	0.00895	9.18650
ΔD	0.02000	Rectangular	B	∞	0.01155	15.29490
Combined standard uncertainty (uc)						0.02953
Effective degrees of freedom (Veff)						302,340.22690
Coverage factor (k)						1.96000
Expanded uncertainty (U)						0.05787

Ra = (0.78 ± 0.06) μm, for a coverage factor k of 1.96 and a confidence of 95%

The measurement results found were Ra = (0.73 ± 0.08) μm for a coverage factor k of 1.96 and a confidence of 95% for the sample ground with 2 passes of ae = 15 μm; and Ra = (0.78 ± 0.06) μm for a coverage factor k of 1.96 and a confidence of 95% for the sample ground with 1 pass of ae = 30 μm. In this case, between the two conditions analyzed, the use of 1 pass of ae = 30 μm is the most appropriate condition for grinding gray cast iron, since the measurement uncertainty was lower and, consequently, the accuracy of the results was higher. In general, the greater the uncertainty associated with roughness, the more the grinding process contributes to the generation of macrogeometric deviations. Furthermore, all roughness values remained below 1.6 μm, which is consistent with roughness acceptable for the grinding process with semi-automatic grinding process machine, like the one employed in this work.

The values obtained for expanded uncertainty were very close to each other and are consistent with other results reported in the literature. Abrão et al. [7] performed grinding tests on 316 UF stainless steel with an aluminum oxide grinding wheel under different cutting conditions, varying the values of ae (15 μm and 30 μm) and obtained expanded uncertainty values of 0.05 μm, 0.06 μm and 0.09 μm. When analyzing the calculations for measurement uncertainty is clear that the variability of readings was the input variable with the greatest contribution to the final Ra roughness measurement uncertainty (approximately 87.06% and 71.69% for two passes of ae = 15 μm and one pass of ae = 30 μm, respectively).

4 Conclusions

The following conclusions can be drawn from this study of measurement uncertainty regarding the roughness parameter Ra of a FC 250 Gy cast iron sample after peripheral surface grinding with a silicon carbide grinding wheel:

- The use of the highest radial depth of cut value (a single pass of ae = 30 μm) resulted in a higher average Ra roughness value, by about 6.4% higher than that obtained after grinding with the lowest ae (performed with two passes of ae = 15 μm).
- The measurement results obtained roughness values after grinding with 2 passes of ae = 15 μm and 1 pass of ae = 30 μm were Ra = (0.73 ± 0.08) μm and Ra = (0.78 ± 0.06) μm, respectively. considering a coverage factor k of 1.96 and a confidence of 95% for both.
- The grinding strategy of using 1 pass with radial depth of cut ae of 30 μm was found to be the most appropriated in grinding of the FC 250 Gy cast iron because the smallest the dispersion of values (measurement uncertainty) obtained under this condition.
- The variability of the readings associated with the Ra values generated the greatest contribution to the final uncertainty of the Ra measurement, being 87.06% and 71.69% for two passes of ae = 15 μm and one pass of ae = 30 μm respectively.

Acknowledgements The authors would like to thank CAPES for their financial support, as well as the Federal University of Uberlândia (UFU) and the Laboratory of Teaching and Research in Machining (LEPU) and the Traditional Machining Laboratory (LUC), both from the School of Mechanical Engineering (FEMEC). Amanda Souza Abrão and Ana Caroline Teixeira Messias would also like to thank CAPES and the Post-Graduate Program in Mechanical Engineering at UFU for the scholarship. The authors also thanks Blaser® Swisslube and Saint-Gobain® Abrasives for providing the cutting fluid and the grinding wheel, respectively. Rosemar Batista da Silva is grateful to CNPq for the Research Productivity scholarship process no. 312274/2022-0 and FAPEMIG—Public Call 002/2018—PPM XII—Process no.: PPM-00492-18.

References

1. Alves MCS (2007) Influence of cooling-lubrication methods on the grinding of ABNT 4340 steel using a CBN grinding wheel. Thesis (Ph.D.)—Doctoral Program in Materials Science and Technology, São Paulo State University (UNESP), Bauru
2. Machado AR, Abrão AM, Silva MB, Coelho RT (2015) Theory of machining of materials, 3rd edn. Blucher Publishing, São Paulo
3. INMETRO (2012) Evaluation of measurement data—guide to the expression of uncertainty in measurement—GUM 2008. Rio de Janeiro, Brazil
4. ISO (2017) ISO/IEC 17025:2017 – General requirements for the competence of testing and calibration laboratories
5. Tupy SA (2014) Cast profiles—technical catalog
6. Lima MLS (2016) Surface grinding of nodular, vermicular, and gray cast irons under various cutting conditions. Dissertation (Master's in Mechanical Engineering)—Federal University of Uberlândia, Uberlândia
7. Abrão AS, Abrão BS, Oliveira MV, Silva RB, Nascimento CG, Arencibia RV (2023) Calculation of measurement uncertainty associated with Ra on interrupted surface of 316 UF stainless steel after grinding. In: Proceedings of the 25th machining colloquium (XXV Colóquio de Usinagem), Brusque (SC), UNIFEBE

Influence of the Reference Position on the Gearing Quality of a Forming Tool

Elliton Ostroski Jaguszeski, Sara Diaz Builes, Fernando Moreira Bordin, Clídio Hort Filho, and Fabio Antonio Xavier

Abstract The requirement for higher precision gears is a constant in various manufacturing processes due to the necessity to improve the performance and efficiency of mechanical systems. Cold forming is one of such processes, especially for the manufacture of precision pinions with complex geometries. Misalignments between axes in the forming system (die and nesting assembly) can generate severe errors, mainly when the required dimensional and shape deviations are of some micrometers. In this context, this study investigates the influence of the reference axis on the measurement of geometric and dimensional deviations of dies used in the cold forming process for the manufacturing of spur gears. The die package tool was analyzed, consisting in the Shrink ring, the upper die and the forming die. The ISO 1328-1/2 standard was used to select the specific parameters to describe the gearing quality of the dies and the DIN 3962 standard was used for the quality classification. The evaluation was done by focusing on the isolated deviations: profile, helix, single and total cumulative pitch, and radial runout. The results showed that the choice of reference axis (top die or gear die) significantly influences some measured deviations. For profile and flank line deviations there was no significant difference, while for pitch and radial run-out deviations significant differences were observed. The statistical analysis of variance (ANOVA) confirmed run-out errors higher than 10 μm, presenting significant influence on the measurement results. This study highlights the importance of choosing the reference axis for dimensional measurement of dies with complex geometries, which is crucial to guaranteeing the desired gear quality produced by cold forming.

Keywords Cold forming · Spur gears · Geometric deviations · Dimensional analysis · Gear class

E. O. Jaguszeski (✉) · S. D. Builes · F. M. Bordin · F. A. Xavier
Laboratory of Engineering Precision, Department of Mechanical Engineering, Federal University of Santa Catarina, Florianópolis, SC, Brazil
e-mail: elliton.ostroski@grad.ufsc.br

C. H. Filho
Zen Metalúrgica S.A, Brusque, SC, Brazil

M. C. Gomes et al. (eds.), *Innovations in Brazilian Machining*, Lecture Notes in Mechanical Engineering, https://doi.org/10.1007/978-3-031-92819-2_9

1 Introduction

The requirement to produce high precision gears is constant in various manufacturing processes, due to the need to improve the performance and efficiency of mechanical systems. Cold forming is one of such processes and stands out especially for the manufacturing of precision pinions with complex geometries.

Forming is a mechanical process that uses high pressures to deform a material in a die, passing the geometry of the die on to the material to be forged [1]. Cold forming is carried out at temperatures below the material's recrystallization temperature, which limits the process to forging carbon steels [2]. This process can produce parts with high precision, reaching IT 7 to IT 11 values [3]. Another advantage of this process is that, due to the hardening generated by the high rate of plastic deformation, it is possible to increase the surface hardness of the component and improve its mechanical properties [4].

According to Ferreira and Guerra [5], producing high-precision technical parts and components in production processes, that are subject to a wide variety of factors, requires large investments in the final inspection. Due to the complexity of designing and manufacturing high-quality gears, there is a huge need to invest resources in the dimensional and geometric control of these parts. Failure to achieve dimensional and geometric tolerances can result in high levels of noise, vibration, friction and damage to the gear set, leading to loss of efficiency and premature damage.

Gear quality is employed to guarantee design requirements following ISO 1328-1/2 [6] standard, which describes measurement strategies and gear quality by means of standardized deviations. This standard covers various types of deviations: profile deviations (F_{α}, $f_{f\alpha}$ and $f_{H\alpha}$), helix deviations (F_{β}, $f_{f\beta}$ and $f_{H\beta}$), pitch deviations (f_{pt} and F_{p}) and runout (F_{r}).

Profile deviations indicate the variation in the shape of the tooth along its profile and are characterized by three parameters: Profile form deviation ($f_{f\alpha}$), which represents the difference between the waviness present in the tooth profile with the theoretical profile shape; Profile slope deviation ($f_{H\alpha}$), which represents the difference in orientation between the actual and theoretical profile; and the total profile deviation (F_{α}) obtained by the ratio of the $f_{H\alpha}$ and $f_{f\alpha}$ values.

The helix deviations reflect the variations in the axial direction of the tooth and, like the profile deviation, also have three parameters that characterize it: helix form deviation ($f_{f\beta}$), where it represents the waviness of the flank surface; helix slope deviation ($f_{H\beta}$), which represents the deviation of the perpendicularity of the flank with the profile surface of the tooth (straight teeth); and total helix deviation (F_{β}) which is obtained from the ratio between $f_{f\beta}$ and $f_{H\beta}$.

The pitch deviations measure the regularity of the pitch between the teeth individually and together, and are also divided into three parameters: single pitch deviation (f_{p}), represents the difference from the actual pitch to the pitch expected by the gear; sector pitch deviation (F_{pk}), can be said to be the sum of the individual pitches of a given region of teeth; and total cumulative pitch deviation (F_{p}), which indicates the largest cumulative pitch deviation found in the gear.

The runout (F_r), on the other hand, is the value resulting from the difference between the maximum and minimum distance of the top of the tooth in relation to the axis of rotation of the gear. In high-precision forming dies, the shrink ring system is applied to increase the mechanical strength and reduce the elastic deformations of the gear die. Shrink ring consists of generating compressive stress in the gear die by means of an enclosure system with the aim of reducing the level of tensile stresses generated during the cold forming process [7].

Based on the topics presented, this paper describes a case study that seeks to investigate the influence of the reference axis on profile and form deviations (evaluation of each quality parameter), trying to assess the degree of relevance of the alignment between dies within an enclosed assembly.

2 Materials and Methods

2.1 The Shrink Ring System and Forming Die

In this work, a shrink ring system was used, as shown in Fig. 1, consisting of two main components: the forming die, which is responsible for the external geometry of the product; and the upper die, which guides the punch and ensures its centralization. Therefore, the homogeneous expansion of the material (blank) and the centralization of the internal and external geometry of the product are ensured through the alignment of these dies, making it essential to validate the alignment accuracy between the components of the shrink ring assembly for the execution of the forming process.

For the deviation analysis, two shrink ring systems were employed. Both systems have: a sintered tool steel die with gear geometry each, used as a tool in a cold-forming process of pinions. Each die has a specific geometry which, in order to maintain the integrity and precision of the cold-forming process, requires double shrinking as illustrated in Fig. 1.

Due to the dimensional deviation of the forming tool, between and outside the hole, the measurement of the gear geometry is carried out with the complete assembly mounted, which is a requirement for achieving the precision and quality of the formed pinion.

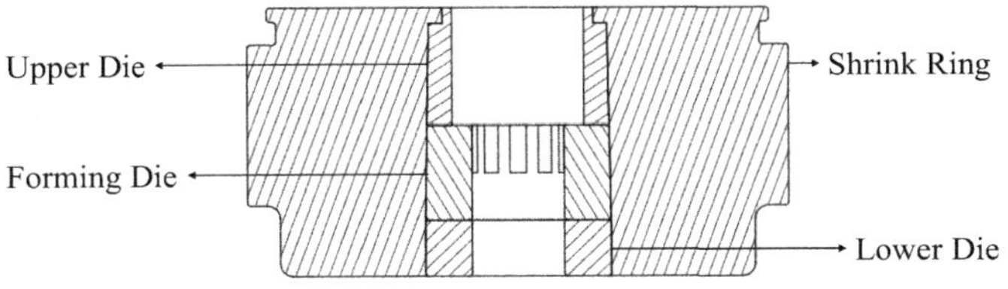

Fig. 1 Forming die shrink ring assembly

By itself, the forming die undergoes different manufacturing stages until it reaches its final geometry. These stages include turning, heat treatment (quenching and tempering), electro-discharge machining (EDM), stress relieving, mechanical polishing with abrasive media, grinding and surface treatment with physical vapor deposition (PVD) coating. More specifically, gear geometry is defined through the die-sinking EDM process, carried out in two phases: roughing and finishing, with the aim of guaranteeing the dimensional quality of the gear profile.

In the forming process, the upper die serves as a guide for the path of the punch that will perform the compressive force on the blank to slide into the gear/forming die. It is, therefore, crucial for the tool manufacturing process to ensure alignment between dies.

2.2 Gear Quality Evaluation

The ISO 1328-1/2 [6] standard is used to specify the gearing quality of the forming die. This standard refers to the control of geometric and dimensional tolerances of spur and helical cylindrical gears and describes the nomenclature of these tolerances [8]. This standard describes two types of deviation: isolated deviations (analytical evaluation) and overall deviations (functional evaluation). For the sake of simplicity, only isolated deviations were assessed:

(a) Profile deviation ($F_{\alpha}, f_{f\alpha}$ and $f_{H\alpha}$);
(b) Helix deviation ($F_{\beta}, f_{f\beta}$ and $f_{H\beta}$);
(c) Single pitch deviation (f_{pt}) and total cumulative pitch deviation (F_p); and
(d) Runout (F_r).

In order to assess the gear quality/grade (Q_n), the individual and overall deviations must be within specific error values, specified by DIN 3962 [9] standard. Lower Q_n values indicate parts with higher gearing quality (smaller tolerances). This standard gives the limit values for each quality class as shown in Table 1, selected according to the gear modulus and diameter, 2.117 and 27.65 mm, respectively in this case.

2.3 Gear Measurement

The specimen was measured in a controlled environment following the criteria and indications according to ISO 17025: 2017 [10]. A ZEISS PRIMO ultra Coordinate Measuring Machine (CMM) with length accuracy of (0.6 + L/500) [μm] was used. For the measurement process, the ZEISS CALYPSO software was used together with GEAR PRO, with an overall uncertainty estimated at 2 μm, considering the contributions of the environment, the equipment and the measurement method.

In order to analyze the influence of the runout between the upper and the forming dies on the result of the gearing quality, two forming dies were measured. These

Table 1 Deviation limit values for each quality class in micrometer [μm]

Gear quality class index	Q_n	1	2	3	4	5	6	7	8	9	10	11	12
Total profile deviation	F_α	2	3	4	5	7	10	14	20	28	45	71	110
Profile form deviation	$f_{f\alpha}$	1.5	2	3	4	6	8	11	16	22	36	56	90
Profile slope deviation	$f_{H\alpha}$	1	1.5	2	3	4.5	6	9	12	18	28	45	71
Total helix deviation	F_β	2.5	3.5	4.5	5.5	7	9	13	18	28	45	71	110
Helix form deviation	$f_{f\beta}$	1.5	2.5	3	3.5	4.5	5.5	7	9	14	25	40	63
Helix slope deviation	$f_{H\beta}$	2	2.5	3	4	6	8	11	16	25	36	56	90
Single pitch deviation	f_{pt}	1	2	2.5	3.5	5	7	10	14	20	32	50	80
Total cumulative pitch deviation	F_p	4	6	8	11	16	20	28	40	56	90	140	250
Runout	F_r	3	4	5.5	8	11	16	22	32	45	63	90	125

Source DIN 3962 [9]

dies underwent the same manufacturing process, differing only in the use of a tool electrode wear protection parameter during the EDM process. These specimens were named Die 1 and 2.

Due to the process fitting conditions, each set of shrink ring system was measured with different probe size and scan lengths, as depicted in Table 2.

It is worth noting that the pitch deviations do not have speed and scan length values, since these deviations are measured using probing points. In addition, for the parameters describing gear quality, a reference axis simulating the gear rotation

Table 2 Measurement parameters for Die 1 and 2 shrink ring systems

Proof body	Deviations	Probe diameter (mm)	Scanning speed (mm/s)	Scan length (mm)
Die 1	$F_\alpha, f_{f\alpha}$ and $f_{H\alpha}$	1.0	0.5	6.03
	$F_\beta, f_{f\beta}$ and $f_{H\beta}$	1.0	2.0	5.50
	F_p and f_{pt}	1.0	–	–
Die 2	$F_\alpha, f_{f\alpha}$ and $f_{H\alpha}$	0.8	0.5	5.96
	$F_\beta, f_{f\beta}$ and $f_{H\beta}$	0.8	2.0	5.50
	F_p and f_{pt}	0.8	–	–

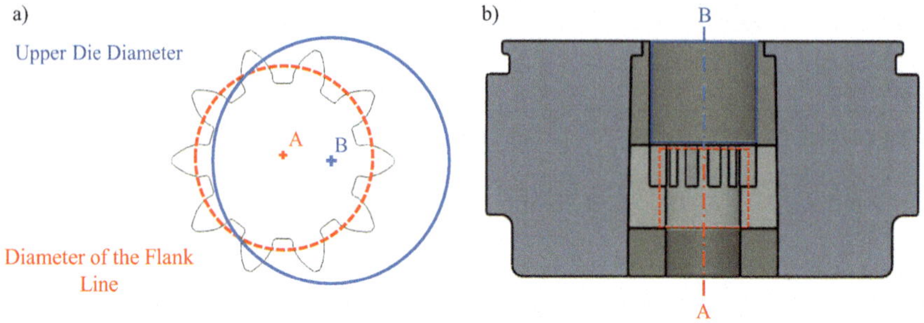

Fig. 2 Schematic representation of the beating error between the upper die and the forming die (**a**) and frontal sectional view of the curling assembly (**b**)

system is used. Therefore, in order to investigate the influence of this axis selection on the measurement of an encased assembly, each system was measured twice: once using the reference axis based on the internal diameter of the upper die and once on the flank diameter of the gearing/forming die.

To understand the difference between the two axes, the runout between the two dies was measured using the software ZEISS CALYPSO which provides the radial runout illustrated in Fig. 2.

In Fig. 2 the distance between point A and B is the value of the dislocation between the reference axes which, in this study, had runout magnitudes of 7.50 μm for Die 1 and 14.0 μm for Die 2.

In order to understand how the concentricity between the dies affects the results and the quality classification of the tool, a comparative analysis of the results was carried out. The geometric deviations of the profile and helix were measured on each die tooth. For this analysis, the mean and standard deviation were calculated, and graphs were plotted for each parameter. In order to investigate statistical representativeness, analysis of variance (p-value) was carried out, with a significance (α) value of 0.05 adopted. The p-value was used to determine whether there was a significant difference between the two measurement processes carried out on each set.

All deviations were measured on both the right and left sides of the gear die tooth, except for runout (F_r).

3 Results and Discussion

The graphs include information on the class or quality (Q_n) to which the deviation belongs, according to the data listed in Tab. 1 of the DIN 3962 [9] standard. Both Fig. 3 and 4 show the results obtained from Die 1 and Die 2 with two measurement processes: using the reference axis from the upper die and gear die, respectively.

There is an overlap of the standard deviation bars, which prevents statistically significant differences from being identified. For the profile deviations ($F_\alpha, f_{f\alpha}$ and

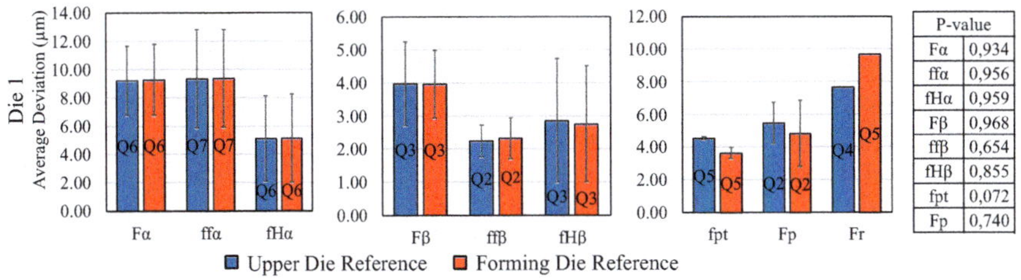

Fig. 3 Die 1 reference comparison considering each selected gear quality parameter

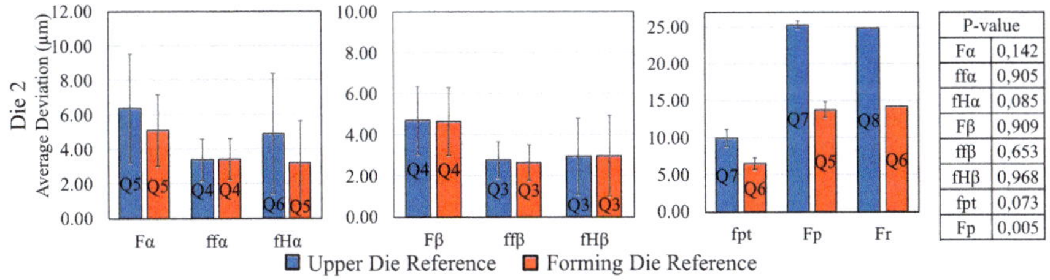

Fig. 4 Die 2 reference comparison considering each selected gear quality parameter

$f_{H\alpha}$), helix deviations (F_β, $f_{f\beta}$ and $f_{H\beta}$) and pitch deviations (f_{pt} and F_p), imposing the reference in one die or another did not result in different gear quality values. This result can also be seen in the obtained p-values, where for none of these deviations the p-value was less than 0.05. For the radial runout F_r, however, there was a change in gear quality from Q4 to Q5 when the measurement reference axis was changed from the top die to the forming die.

This can be best explained from the measurement process: if the upper die is not aligned with the forming die, which has gear geometry, the tooth top will not be in the correct position estimated by the CMM. This lack of alignment results in a probing error consistent with the magnitude of the runout between the dies. This situation ends up jeopardizing the accuracy of the measurement process and, therefore, the assessment of the quality of the gearing. Figure 4 showcases the behavior of the gear quality parameters for Die 2.

When analyzing the data of die 2 in Fig. 4, the profile slope deviation ($f_{H\alpha}$) shows a disparity between the measurements, which is also reflected in the profile deviation total (F_α) result. This behavior is due to the method in which the total profile deviation is obtained: by the ratio of profile form deviation ($f_{f\alpha}$) and profile slope deviation ($f_{H\alpha}$) Eq. (1).

$$F_\alpha = \sqrt{(f_{f\alpha})^2 + (f_{H\alpha})^2} \quad (1)$$

The helix deviations (F_β, $f_{f\beta}$ and $f_{H\beta}$) did not showcase a significant difference between the two samples. However, when looking at the values for pitch deviation

(f_{pt} and F_p) and runout (F_r), highlighted in Fig. 4, there is a significant difference between the average values and the gear class to which they belong. In addition, when analyzing the p-values from the analysis of variance, the p-value for the profile slope deviation ($f_{H\alpha}$) and single pitch deviations (f_{pt}) is close to 0.05. When looking at the total step deviation (F_p), the p-value was less than 0.05, indicating a statistically significant difference between the groups.

The comparison between the measurement procedures, using the reference axis in the upper die and another in the forming die, shows that, although the profile and helix deviations showed similar magnitudes, the pitch deviations and runout showed significant differences, confirmed by the p-values in the ANOVA analysis. When comparing the difference in results for runout of 7.50 μm (Die 1), it can be seen that there is no statistically representative influence, aside from the pitch deviation. On the other hand, when comparing the difference in results for runout of 14.0 μm (Die 2), it was observed that the slope profile and pitch deviations are statistically significant different. As a result, this indicates that deviations in the order of 10 μm could alter the measurement.

4 Conclusion

This case study presented the influence of the choice of reference axis on the results of geometric and dimensional deviations for forming dies.

For a runout error of 7.5 μm (Die 1) and 14.0 μm (Die 2), the profile and flank deviations showed no statistically significant changes. However, for a runout of 14.0 μm (Die 2), the $f_{H\alpha}$ deviation is susceptible to significant influences, as evidenced by a p-value of 0.085.

With regard to the runout and pitch deviations, when the difference between axes is observed with a runout of 14.0 μm (Die 2), there was a significant difference in relation to the gear class described by the values of Q7 and Q6 for (f_{pt}), Q7 and Q5 for (F_p) showing a statically significant difference with a p-value of 0.005, Q8 and Q6 for (F_r). When the runout was 7.5 μm (Die 1) only the runout deviation showed a change in class, being Q4 and Q5 and with an individual step deviation value showing a p-value close to 0.05.

The statistical analysis of variance confirmed that runout errors greater than 10 μm could representatively alter the measurement results. Consequently, there can be changes in the quality of gearing by increasing or decreasing class according to the reference axis. This highlights the importance of choosing the reference axis when measuring forming dies when measuring the curl assembly.

References

1. Gemaque MJA (2004) Approach for metrological problem solving in industry: gear measurement. Master's thesis. Federal University of Santa Catarina, Florianópolis, Santa Catarina, p 132
2. Bôas DV (2009) A contribution to cold forging study of spiral gears. State University of Campinas, Campinas, p 183
3. Hirschvogel M, Dommelem HV (1992) Some application of cold and warm forging. J Mater Process Technol 35:343–356
4. Bôas DV (1999) Cold forging process study of automative parts. Master's thesis. State University of Campinas, Campinas, p 116
5. Ferreira F, Guerra H (2018) The coordinate metrological challenges in dimensional and geometrical specifications control of technical components in the new 4.0 industry paradigm. J Prod Dev 4(1):125–132
6. International Organization for Standadization 1328-1 (2013) Cylindrical gears—ISO system of accuracy, Part 1. Definitions and allowable values of deviations relevant to radial composite deviations and runout information. ISO, Switzerland
7. Silveira FD (2019) Study of the influences of strapping rings on cold extrusion dies of cylindrical spur gears with the 10B22 steel for automative industry. Doctoral dissertation. Federal University of Rio Grande do Sul, Porto Alegre, p 135
8. Júnior LS (2010) Quality assurance system for measuring parts with complex geometry and free-form surfaces using a coordinate measuring machine. Doctoral dissertation. Federal University of Paraíba, Paraíba, p 206
9. Deutsches Institut für Normung 3962-1 (1978) Tolerances for cylindrical gear teeth, Part 1. Tolerances for deviations of individual parameters. DIN, Berlin
10. International Organization for Standadization/IEC 17025 (2017) General requirements for the competence of testing and calibration laboratories. ISO, Switzerland

Measurement Uncertainty Calculation Associated with Ra Roughness of Titanium Alloy After Grinding

João Vitor Antonieti Reis, Josenilton dos Santos Lopes, Alcione dos Reis, Amauri Hassui, Rosemar Batista da Silva, and Mayara Fernanda Pereira

Abstract Titanium alloys are materials commonly used in the manufacture of components in aeronautical industry due to a unique combination of properties such as high mechanical strength, low density, high corrosion and oxidation resistance. However, titanium alloys machinability is compromised due to its low thermal conductivity and high chemical reactivity with tool materials, especially with ceramics (major materials applied in in grinding process). Therefore, is essential to study and define the most suitable cutting conditions that includes of dressing operation, that directly affects the component surface quality. In this sense, the calculation of measurement uncertainty associated with Ra roughness parameter of the Ti-6Al-4V alloy was developed after dressing the grinding wheel under different conditions. The expression was based on proposed methodology by Guide to the expression of uncertainty in measurement. Experimental tests were developed with a silicon carbide grinding wheel conditioned under two dressing conditions: an aggressive overlap ratio $U_d = 3$, and a smooth overlap ratio condition, $U_d = 12$. Cutting speed (30 m/s), workspeed 7.5 m/min and radial depth of cut (30 μm) were kept constant in all the tests. Surface roughness (Ra parameter) and measurement uncertainty associated with it were the output variables analyzed. The results showed a reduction of 32.9% in Ra parameter due to a change in grinding wheel conditioning from an aggressive condition ($U_d = 3$) to a smooth one ($U_d = 12$). From analyzed results in terms of measurement uncertainty, it was possible to identify that under $U_d = 12$ dressing condition the lower Ra roughness results were obtained.

Keywords Grinding · Measurement uncertainty · Titanium alloys · Dressing · Ra roughness parameter

---

J. V. A. Reis · J. dos Santos Lopes (✉) · A. dos Reis · R. B. da Silva · M. F. Pereira
Faculty of Mechanical Engineering, Federal University of Uberlândia, Uberlândia, MG, Brazil
e-mail: josenilton.lopes@ufu.br

A. Hassui
Faculty of Mechanical Engineering, State University of Campinas, Campinas, SP, Brazil

M. C. Gomes et al. (eds.), *Innovations in Brazilian Machining*, Lecture Notes in Mechanical Engineering, https://doi.org/10.1007/978-3-031-92819-2_10

1 Introduction

Among various types of titanium alloys, Ti-6Al-4V alloy stands out, especially in the manufacture of aircraft engine blades, as well as fuselage components because of its unique properties combination such as low density (4.43 $g \cdot cm^{-3}$), high tensile strength (895–1170 MPa) at extreme temperatures, high corrosion and oxidation resistance, resistance to fatigue (400–500 MPa) [1].

However, titanium alloys are considered difficult to grind materials due to their high mechanical resistance, even at high temperatures, high chemical reactivity with ceramic materials (major materials of grinding wheels), as well as low thermal conductivity (6.6 $W \cdot m \cdot K^{-1}$), which causes a significant portion of the heat generated by machining to be retained in the workpiece during grinding [2]. Throughout grinding, when the abrasive grits are in contact with the surface of the titanium alloy, they become highly chemically reactive with the titanium element, leading to deterioration of surface finish, high temperatures, and increase in heat generation on the grinding wheel-workpiece interface. So, under these conditions the workpiece will be more susceptible to thermal damages, thereby adversely affecting its functionality [3].

In this sense, in order to improve machinability of the Ti-6Al-4V alloy becomes necessary to test a wider range of cutting parameters in comparison with steels, for instance, and evaluate the most appropriate cutting parameters in order to ensure the projects requirements in terms of surface finish and interchangeability [4, 5].

However, throughout the machining process, abrasive grits frequently wear out and lose their cutting edge, making it essential to sharpen this grinding wheel, through a process called dressing. This operation aims to recondition the worn grinding wheel by sharpening the edges of the abrasives and exposing new and friable edges with the aid of a dresser [6]. So, the surface roughness and texture of the workpiece directly depend on the dressing operation.

Especially in difficult-to-machine alloys, such as Ti-6Al-4V alloy, for instance, measuring surface roughness is essential for evaluating the effects of grinding and dressing parameters on the surface quality of the workpiece material [3, 7]. Roughness is defined as a group of microgeometrical irregularities, characterized by protuberances and indentations on a surface [8, 9]. There are different roughness parameters that allow a surface to be characterized; however, the most popular parameter is Ra roughness surface parameter, which is defined as the arithmetic mean of the absolute values of the ordinates Z(x), along the sampling length, Eq. (1), where "Z(x)" is the ordinate and "l" is the number of ordinates considered [8, 9].

$$Ra = (1/l) \cdot \int |Z(x)| dx \qquad (1)$$

Accordingly, the analysis and calculation of measurement uncertainty is essential, given that this is defined as a metrological parameter that allows characterizing the dispersion of values attributed to a measurand (quantity to be measured), based on applied information [10].

In this way, the measured data contributes to the traceability of the measurement result, according to the standard [11]. In addition, there is the fact that the evaluation and declaration of the measurement uncertainty associated with the Ra roughness parameter will ensure a reliable comparison between the measurement results achieved and those reported elsewhere in the literature or reference values. Therefore, the use of the methodology proposed by the Guide for expressing measurement uncertainty can be used [10, 12].

In a work on grinding process developed by [13] the authors analyzed the influence of different dressing depth conditions (5, 10, 15, 20, and 25 μm), using a single diamond tip dresser. On the Ra roughness parameter of the Ti-6Al-4V alloy. It was used a white aluminum oxide grinding wheel (AA60K5V8) under dry condition. The authors verified that the increase in depth of dressing cut, up to the condition of 20 μm, provided an increase in Ra roughness from 0.7 to 0.8 μm. However, under the most severe condition of dressing depth, 25 μm, there was a significant drop in values, on average 0.6 μm [13]. The authors attributed this behavior to the geometry of the abrasives after dressing which, under the most aggressive condition, became more sharpened, due to micro and macro effects, thus, the surface of the samples Ti-6Al-4V exhibited a better surface finish, in terms of Ra roughness.

Therefore, the present work aimed to analyze the effects of the dressing operation on the surface finish of the Ti-6Al-4V alloy, by calculating the measurement uncertainty associated with roughness. In this study, two different dressing conditions were used: overlap ratio of $U_d = 3$ (severe condition), $U_d = 12$ (smooth condition) and, as an output variable, the Ra surface roughness parameter was measured and evaluated. And, with the analysis of measurement uncertainty, it was possible to identify the most suitable condition for grinding the titanium alloy.

2 Experimental Methodology

Grinding tests were developed in a semi-automatic peripheral surface grinder model P36, with a z-axis resolution of 5 μm, rotation on the main axis of 2400 rpm, and a useful power of 2.2 kW (3 HP) from the manufacturer MELLO S.A.

The workpiece material was the Ti-6Al-4V alloy with dimensions: 37 mm (length) × 15 mm (height) × 23 mm (width). In order to ensure the proper positioning of the parts, each sample was clamped in a precision vise and positioned on the machine table. The workspeed of 7.5 m/min was kept constant. Furthermore, all grinding trials were carried out with a semi-synthetic cutting fluid delivered under flood conditions and at a flow rate of 9 L/min.

A silicon carbide grinding wheel with C46K5V5 designation and dimensions 304.8 mm (external diameter) × 25.4 mm (width) × 76.2 mm (internal diameter) was employed. The cutting speed of 38.3 m/s was kept constant. Prior to each grinding test, the silicon carbide grinding wheel was subjected to a dressing operation with aid

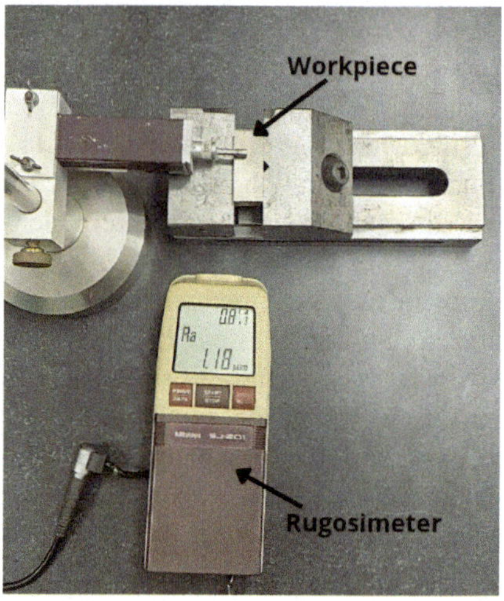

Fig. 1 Roughness measurement setup on Ti-6Al-4V samples, after grinding

of a fliesen-type dresser, at a depth of dressing cut equal to ad = 40 μm, and in two different overlap ratio conditions, $U_d = 3$ (severe condition) and $U_d = 12$ (smooth condition). After dressing, experimental grinding tests were carried out using 5 passes of the grinding wheel on the workpiece with of 30 μm each one.

After the grinding tests on the titanium samples, the surface roughness was measured in terms of the Ra parameter, so that the effects of different overlap ratio conditions on the surface finish. All measurements were carried out in the Metrology Teaching Laboratory (LEMETRO–UFU), at a controlled temperature (20 ± 1) °C, using a portable rugosimeter, (Surftest model SJ201P, from Mitutoyo manufacturer, with 0.01 μm resolution, and tip radius of 2 μm (Fig. 1).

The selected sampling length (cut-off) was 0.8 mm, with an evaluation length of 4.0 mm. The roughness meter probe was positioned so that it moved perpendicularly to the direction of the grooves generated by grinding wheel in 5 distinct and equally spaced regions. After surface roughness measurements, an analysis and calculation of the measurement uncertainty associated with roughness was carried out, according to the method proposed by the Guide for expressing measurement uncertainty (GUM) and with the following the steps [10, 12]:

1. Definition of the measurand or output variable (Ra);
2. Identification of input variables that may affect the measurement of the output variable;
3. Mathematical model of the measurand as a function of all input variables;
4. Calculation of the standard uncertainty associated with each input variable (u);
5. Calculation of the combined standard uncertainty regarding the output variable (uc);
6. Calculation of effective degrees of freedom;
7. Assessment of expanded uncertainty regarding roughness (U);
8. Expression of measurement result.

Even though the calibration certificate of surface roughness tester was not provided, the analysis of measurement uncertainty was carried out based on five input variables being detected: the variability of the readings (x), finite resolution of the rugosimeter (R), uncertainty associated with surface roughness tester calibration (C), uncertainty associated with the probe tip radius (Rp) and uncertainty associated with deformation of workpiece material (D) [14]. Thus, the mathematical model for Ra roughness is, Eq. (2):

$$Ra = x + \Delta R + \Delta C + \Delta Rp + \Delta D \tag{2}$$

The uncertainty propagation law was applied to Eq. (2) in order to determine the equation relating to combined standard uncertainty (u_c), Eq. (3):

$$u_c^2(Ra) = u^2(x) + u^2(\Delta R) + u^2(\Delta C) + u^2(\Delta Rp) + u^2(\Delta D) \tag{3}$$

Table 1 lists the quantities (input variables) and their estimates, the type of uncertainty assessment (T.A), the probability distribution (D.P), the degrees of freedom (G.L) as well as the coefficients of sensitivity (C.S) and calculations for standard uncertainty. Where "s" is the sample standard deviation, "Rrug" is the resolution of the roughness meter, "Prug" is the average value of the roughness parameter, and "D" is the elastic deformation of the alloy during the perpendicular movement of the probe, equivalent at 20 nm [14].

Once that the standard uncertainty values (u) are calculated, the next step is to calculate the combined standard uncertainty (u_c) by the Eq. (3), and subsequently, the effective degrees of freedom (υ_{eff}) by the Eq. (4), as well as the expanded uncertainty (U) using the Eq. (5). Finally, the percentage contribution of each of the input variables to the expanded uncertainty was also calculated by the Eq. (6).

$$\vartheta_{eff} = u_c^4(y) / \sum\left(\left(u^4(y_i)\right)/v_i\right) \tag{4}$$

$$U(Ra) = k \cdot u_c(Ra) \tag{5}$$

$$Cont(\%) = ((u(x_i) \cdot C_i))^2 / (u_c(y))^2 \tag{6}$$

Table 1 Standard uncertainties information

Quantity	Estimation	P.D	T.E	D.F	C.S	Standard uncertainty
x	x	A	t-student	n – 1	1	$u(x) = s/\sqrt{3}$
ΔR	Rrug	B	Rectangular	∞	1	$u(\Delta R) = Rrug/2\sqrt{3}$
ΔC	Rrug	B	Normal	∞	1	$u(\Delta C) = Rrug/k(C)$
ΔRp	0.02 · Prug	B	Rectangular	∞	1	$u(\Delta Rp) = 0.02 \cdot Prug/\sqrt{3}$
ΔD	D	B	Rectangular	∞	1	$u(\Delta D) = D/\sqrt{3}$

3 Results

From the Fig. 2 can be seen that roughness measurements and calculation of measurement uncertainty indicated a reduction of 32.90% in the roughness parameter Ra when grinding overlap ratio was changed from a roughing condition ($U_d = 3$) to a finishing one ($U_d = 12$). When grinding with $U_d = 3$ dressing condition, the measurement result in terms of arithmetic average and expanded uncertainty was Ra = (1.55 ± 0.14) μm, to a k = 1.96 coverage factor and a reliability of 95%. On the other hand, when grinder under $U_d = 12$ dressing condition, the measurement result was Ra = (1.04 ± 0.19) μm, to a k = 1.96 coverage factor, and a reliability of 95%, Fig. 2.

In Tables 2 and 3 are presented the necessary information to express the measurement uncertainty associated with roughness Ra values, under $U_d = 3$, and $U_d = 12$ conditions, respectively. This analysis ensures traceability in the results as well as a reliable comparison with data available in a specific literature.

According to Klocke [6] the kinematics of the dressing operation generates a helical profile on the cutting surface of the grinding wheel, which encompasses macro and micro effects, depending on the texture generated on the grinding wheel. When an aggressive overlap ratio $U_d = 3$ is selected, the grinding wheel generated texture ensures a roughing condition for grinding. Thus, there will be a greater contact area between abrasives and workpiece, thereby generating higher material removal rate. As a consequence, the surface finish and texture of the workpiece are characterized by deeper grooves and intense plastic deformation, thereby promoting higher values of surface roughness. However, under a smooth overlap ratio condition, such as $U_d = 12$, improved finish texture is generated on the grinding wheel with low material removal rate, leading to low values of surface roughness values, in comparison to aggressive dressing conditions [6].

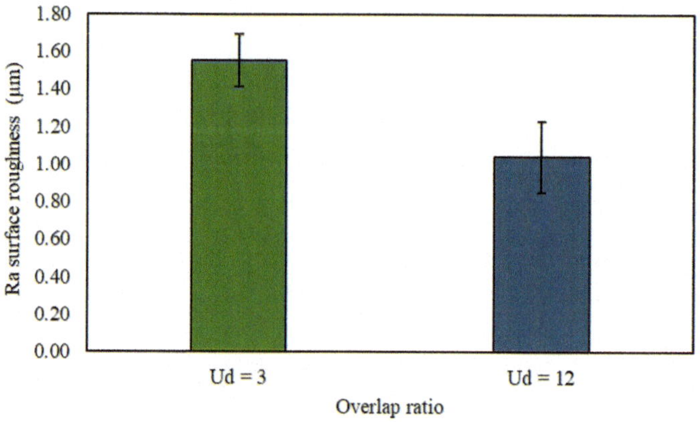

Fig. 2 Ra surface roughness values in Ti-6Al-4V alloy after grinding with a SiC wheel conditioned under $U_d = 3$, and $U_d = 12$ overlap ratio conditions

Table 2 Uncertainty associated with Ra surface roughness parameter, under Ud = 3 condition

Quantity	Estimation	P.D	T.E	D.F	Standard uncertainty	Contribution
x	1.55400	A	t-student	4	0.06765	90.35165
ΔR	0.01000	B	Rectangular	∞	0.00289	0.16454
ΔC	0.01000	B	Normal	∞	0.00500	0.49362
ΔRp	0.03108	B	Rectangular	∞	0.01794	6.35757
ΔD	0.02000	B	Rectangular	∞	0.01155	2.63262
Combined standard uncertainty (uc)						0.07117
Effective degrees of freedom (V_{eff})						13,594.5
Coverage factor (k)						1.96000
Expanded uncertainty (U)						0.13949

Table 3 Uncertainty associated with Ra surface roughness parameter, under Ud = 12 condition

Quantity	Estimation	P.D	T.E	D.F	Standard uncertainty	Contribution
x	1.04400	A	t-student	4	0.09760	96.82871
ΔR	0.01000	B	Rectangular	∞	0.00289	0.08471
ΔC	0.01000	B	Normal	∞	0.00500	0.25412
ΔRp	0.02088	B	Rectangular	∞	0.01206	1.47718
ΔD	0.02000	B	Rectangular	∞	0.01155	1.35529
Combined standard uncertainty (uc)						0.09919
Effective degrees of freedom (V_{eff})						4372.12
Coverage factor (k)						1.96000
Expanded uncertainty (U)						0.19441

A similar behavior was observed by [5] on Ti-6Al-4V surfaces after grinding with a C46K5V5 silicon carbide grinding wheel. The authors observed that grinding wheel generated texture, by the dressing operation, was printed on ground surface of Ti-6Al-4V alloy, and Ra roughness parameter is the most directly influenced by the action of abrasive grains on workpiece surfaces. [5] observed that a change in overlap ratio from $U_d = 3$ to $U_d = 12$ could reach up to 53.74%.

From Tables 1 and 2 can be observed that variability of measured values was the quantity that most contributed to measurement uncertainty, 90.35% under $U_d = 3$ condition, and 96.82% under $U_d = 12$ one.

In this way, by expressing the measurement uncertainty associated with the roughness parameter Ra, it is possible to identify which input variables influence the roughness values, the percentage of contribution of each of them to the result and, mainly, verify reliably that the change in the grinding wheel conditioning generated a significant reduction in the measured parameter. This measurement uncertainty analysis indicates that, in terms of roughness, the most suitable condition for grinding the Ti-6Al-4V alloy is $U_d = 12$.

4 Conclusions

From the experimental grinding tests of Ti-6Al-4V and analysis of the results, the following conclusions can be drawn:

- A change in grinding wheel dressing from an aggressive condition ($U_d = 3$) to a smooth one ($U_d = 12$) lead to 32.90% reduction in roughness Ra.
- Measurement uncertainty analysis showed that Ra roughness parameter reduced from (1.55 ± 0.14) μm, for a k = 1.96 coverage factor, and a reliability of 95% ($U_d = 3$ overlap ratio condition)) to (1.04 ± 0.19) μm, for a k = 1.96 coverage factor, and a reliability of 95% ($U_d = 12$ overlap ratio condition).
- The variability of measured values was the quantity with the greatest contribution to measurement uncertainty.
- From the measurement uncertainty analysis, it was possible to identify with metrological reliability that the most suitable condition for Ti-6Al-4V alloy grinding is $U_d = 12$.

Acknowledgements The authors would like to thank the governmental agencies: The Federal Agency for the Support and Improvement of Higher Education (CAPES), the National Council for Scientific and Technological Development (CNPq), Minas Gerais State Research Foundation (FAPEMIG), and the Graduate Program in Mechanical Engineering at the Federal University of Uberlândia for financial support. We also thank the organizing committee of the XXVI Machining Colloquium

References

1. Leyens C, Peters M (2003) Titanium and titanium alloys—fundamentals and applications, 1st edn. Wiley–VHC, Weinheim
2. Anand N, Kumar AS, Paul S (2019) Effect of cutting fluids applied in MQCL mode on machinability of Ti-6Al-4V. J Manuf Process 43:154–163
3. Malkin S, Guo C (2008) Grinding technology—theory and applications of machining with abrasives, 2nd edn. Industrial Press, New York
4. Mukhopadhyay M, Kundu PK, Chatterjee S, Das S (2019) Impact of dressing infeed on SiC wheel for grinding Ti-6Al-4V. Mater Manuf Processes 1:54–60
5. Pereira MF, Abrão BS, Abrão BS, Machado AR, Jackson MJ, Hassui A, Da Silva RB (2023) Influence of grinding wheel conditioning on the grindability of Ti-6Al-4V alloy. Int J Adv Manuf Technol 125:1531–1542
6. Klocke F (2009) Manufacturing process 2—grinding, honing, lapping, 2nd edn. Springer, Aachen
7. Raj RB, Mukhopadhyay M, Kundu KP (2019) Evaluating the grinding ratio and surface quality of Ti-6Al-4V under varying pass count and depth of cut. J Phys Conf Ser 1240:1–8
8. Gadelmawla ES, Kourab MM, Maksoudc TMA, Elewaa IM, Solimand HH (2002) Roughness parameters. J Mater Process Technol 123:133–145
9. ABNT NBR ISO 4287 (2002) Geometrical Product Specifications (GPS) — Surface texture: Profile method — Terms, definitions and surface texture parameters
10. Inmetro (2012) Evaluation of of of of measurement data measurement data measurement data, 3rd edn. Rio de Janeiro

11. ABNT NBR SO/IEC 17025 (2017) General requirements for the competence of testing and calibration laboratories
12. JCGM (2008) Evaluation of measurement data-guide to the expression of uncertainty in measurement, vol 1. Joint Committee for Guides in Metrology, pp 1–134
13. Mukhopadhyay M, Kundu PK (2018) Optimization of dressing infeed of alumina wheel for grinding Ti-6Al-4V. Mater Manuf Processes 33:1453–1458
14. Leach R (2001) The measurement of surface texture using stylus instruments. Measurement Good Practice Guide, National Physical Laboratory, Teddington, Middlesex

Analysis of Surface Roughness in Micromilling AISI 316L Stainless Steel with Variation of Feed per Tooth

Milla Caroline Gomes, Maria Clara Coimbra Gonçalves, and Márcio Bacci da Silva

Abstract The development of micromachining has enabled the production and improvement of components and features on microscale, which are essential in many industries. However, as the scale decreases, phenomena such as ploughing and tool run-out significantly impact product quality, making it crucial to carry out studies to understand the cutting mechanisms on a microscale, which change depending on the characteristics of the process performed. Therefore, the objective of this study is to evaluate how the feed per tooth affects the quality of the machined surface in the micromilling process of AISI 316L austenitic stainless steel. For this purpose, 0.4 mm diameter micromills were used to manufacture microchannels using feed rate of 0.5, 1 and 5 μm/tooth. The machined surfaces of each microchannel were observed using a scanning electron microscope, and their surface roughness was measured. These analyses were conducted to assess the characteristics of the machined surface on each microchannel. According to the experimental results obtained for micromilling AISI 316L with the selected cutting parameters, the average surface roughness (Ra) showed no statistically significant difference between feeds of 0.5 μm/tooth and 1 μm/tooth, both yielding an Ra of approximately 0.04 μm. However, when the feed rate was increased from 1 to 5 μm/tooth, the surface roughness also increased, with the average surface roughness (Ra) rising from 0.04 to 0.15 μm. In the experiments performed, the influence of the scale effect on the surface finish was not observed, as the selected feed per tooth values are higher than the estimated minimum chip thickness (0.27–0.37 μm). Analysis of the micromachined surface images showed that lower feed rates result in a better surface finish, while higher feed rate led to material detachment, particle adhesion, and the formation of internal burrs due to side flow.

M. C. Gomes (✉)
Department of Manufacturing Engineering and Materials, University of Campinas, Campinas, SP, Brazil
e-mail: millagm@unicamp.br

M. C. C. Gonçalves
Mechanical Engineering Department, University of Sheffield, Sheffield, UK

M. B. da Silva
School of Mechanical Engineering, Federal University of Uberlandia, Uberlandia, MG, Brazil

M. C. Gomes et al. (eds.), *Innovations in Brazilian Machining*, Lecture Notes in Mechanical Engineering, https://doi.org/10.1007/978-3-031-92819-2_11

Keywords Micromachining · Micromilling · Roughness · Feed per tooth · Stainless steel

1 Introduction

Micromachining is a manufacturing process used to create three-dimensional geometries in a wide range of materials. Among the micromachining processes, micromilling is defined as the scale-down version of the traditional milling process, where tools up to 1 mm in diameter are used to remove material and give shape to a part [1]. Micromilling is used in sectors such as the medical, dental, aeronautical, and jewellery industries to produce implants, gears, tools, and other microscale components [2]. However, micromachining presents specific challenges related to the reduced scale, such as the ploughing effect and tool run-out [3], especially when machining difficult-to-process materials like AISI 316L stainless steel. Therefore, understanding the details of this process is crucial for comprehending its cutting mechanics, enabling advancements in the field, and opening up a range of possibilities for manufacturing components on a reduced scale.

AISI 316L stainless steel is a material known for its biocompatibility and high corrosion resistance, desirable properties for many applications in the medical and petrochemical fields. In this context, Tamez-Tamez et al. [4] analysed the feasibility of manufacturing microneedles using AISI 316L stainless steel through the micromilling process. Microneedle structures are an emerging technology designed to deliver drugs into human tissue. The authors concluded that the micromilling process was viable for their production, meeting the desired specifications and ensuring the functionality of the microneedles [4]. Despite its wide range of applications, AISI 316L stainless steel is a challenging material to machine due to its low thermal conductivity, which makes it difficult to dissipate heat in the tool-workpiece contact zone. For this reason, low cutting speeds are recommended to mitigate the work hardening that occurs when this material is subjected to mechanical processes [5].

Kajaria et al. [6] investigated the micromilling of 316L stainless steel using the minimum quantity lubrication (MQL) technique. They employed fluid-dynamic models to analyse the behaviour of the fluid around the rotating microtool. The authors concluded that, when applied properly, the minimum quantity lubrication reduces the built-up edge formation and increases tool life by up to 100 times compared to dry micromilling. Later, Gomes et al. [7] carried out an experimental study on the micromilling of the same material, evaluating the influence of the machined length on tool wear, burr formation, and surface roughness. The authors found that the behaviour of tool wear in terms of machined length was similar to that of conventional tools; burr formation in the down-milling side was higher; and burr formation decreased with machined length. Cutting speed was the only parameter that influenced the surface roughness.

In a more recent study, Greco et al. [8] investigated the micromilling process of 316L manufactured by additive manufacturing. The authors investigated the impact of process parameters and build-up direction of the selective laser melting (SLM) process on the micromilling process outcomes. The authors used 50 μm diameter tools with one cutting edge to manufacture slots parallel and perpendicular to the material build-up direction. They observed significant differences when varying the feed direction relative to the material build-up direction, due to the characteristic microstructure of parts made by SLM. Additionally, the authors varied the energy density of the additive manufacturing process for comparison purposes. As a result, the more porous material exhibited higher roughness, whereas the less porous material resulted in greater tool wear due to the increased amount of material removed.

To contribute to these investigations, the present study aims to analyse the influence of feed per tooth on the quality of the machined surface during the micromilling of AISI 316L austenitic stainless steel. For this purpose, experiments were conducted by varying the feed per tooth while keeping other parameters constant. The goal was to compare the effect of feed on the surface roughness (Ra) and on the quality of the machined surface, with the latter evaluated through electron microscopy.

2 Materials and Methods

The workpiece material used was AISI 316L austenitic stainless steel, which is widely applicable in the medical field due to its biocompatibility and high corrosion resistance and other characteristics. Table 1 shows the chemical composition of this material.

The mechanical properties of AISI 316 L are shown in Table 2.

During the tests, microchannels were manufactured using the micromilling process under dry condition. The Minitech Machinery Corporation® CNC Mini-mill/Gx with a maximum spindle speed of 60,000 rpm, three axes, and a positioning resolution of 0.1 μm, was used to perform the experiments. This machine is located on a granite inertial table designed to reduce vibrations from external sources during micromachining.

Table 1 Chemical composition of the AISI 316L stainless steel [9]

Element	C	Si	Mn	P	S	Cr	Ni	Mo
Percentage	≤ 0.03	≤ 1.0	≤ 2.0	≤ 0.045	≤ 0.03	16.00–18.0	10.0–14.0	2.0–3.0

Table 2 AISI 316L mechanical properties [9]

Ultimate tensile strength (Min. MPa)	Yield strength (Min. MPa)	Hardness HB (Max. MPa)
515	205	183

A total of six microchannels were manufactured considering the tests and replicas, three different feed per tooth values were used, while other parameters were kept constant, as shown in Table 3. For each microchannel manufactured, a new micromill was used in order to minimise the influence of tool wear on the comparison of outcomes.

The microtools used were carbide flat end mills coated with (Al, Ti)N, with a cutting diameter of 0.4 mm and two cutting edges, manufactured by Mitsubishi Materials. These micromills have cutting lengths (APMX) of 800 μm, and a shank diameter (DCON) and length (LF) of 4 mm and 40 mm, respectively, as illustrated in Fig. 1.

Figure 2a shows the side view and Fig. 2b the top view images of one of the micro end mills used, obtained using the Hitachi High-Technologies® Scanning Electron Microscope, model TM3000. Figure 2c illustrates the procedure used to measure the cutting-edge radius ($r\varepsilon$) of the micromills. The value of this radius was estimated using the Image J software and it was equal to 1.1 ± 0.1 μm.

For each microchannel manufactured, images of the machined surface were evaluated using the Hitachi High-Technologies® Scanning Electron Microscope, model

Table 3 Design of experiments

Spindle speed (rpm)	Cutting speed (m/min)	Feed per tooth (μm/tooth)	Feed speed (mm/min)	Axial depth of cut (μm)	Radial depth of cut (μm)
10,000	12.6	0.5	10	40	400
		1.0	20		
		5.0	100		

Fig. 1 Micromills geometry

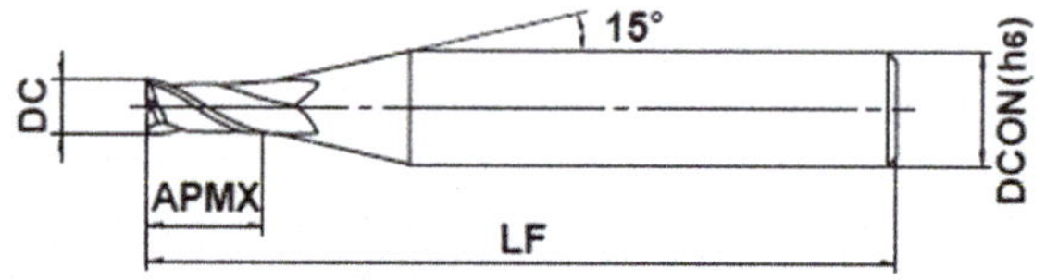

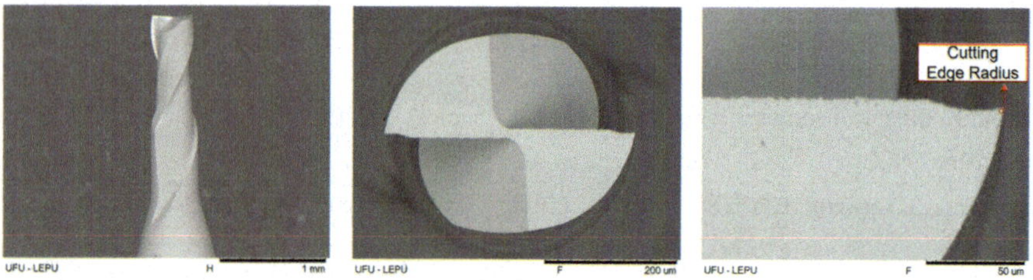

(a) (b) (c)

Fig. 2 Images of the micromills before use, illustrating the procedure for measuring the cutting-edge radius

TM3000, beside, the surface roughness was measured using a Taylor Hobson® profilometer, model Form Talysurf Intra, which has a resolution of 16 nm and a 2 μm diameter diamond tip probe. Measurements were performed perpendicular to the feed marks, in an environment with controlled temperature of around 20 °C. Three measurements were taken in different regions of each microchannel using a Gaussian filter with a cut-off of 0.8 mm and an evaluation length of 4 mm, as specified in the ISO 4288 [10] standard. For analysis, the average of these measurements and the standard deviation were obtained. The parameter analysed was the arithmetic mean deviation (Ra).

3 Results and Discussion

Figure 3 shows the surface roughness (Ra) results as a function of the feed per tooth for the micromilling process analysed. The results presented are the average Ra values obtained from the test and its replica, with standard deviation within a 68.27% confidence interval.

From Fig. 3, a high standard deviation in the obtained Ra results can be observed, with values overlapping for the feeds of 0.5 and 1 μm/tooth. These results indicate that there are no statistical differences in roughness between the feeds of 0.5 and 1 μm/tooth. However, the Ra values differ when the feed is significantly increased from 1 to 5 μm/tooth, where it is evident that the surface roughness increases with the feed per tooth. This behaviour is similar to conventional milling, where a proportionally quadratic increase occurs in the height of the peaks and in the depth of the valleys in the feed marks relative to the feed used [11].

Furthermore, for the feed per tooth values analysed, the influence of size effect on the surface finish of the machined microchannels was not observed, as the selected feed values were greater than the minimum chip thickness, according to the estimation model presented by De Oliveira et al. [12]. According to these authors, the minimum chip thickness varies from 1/4 to 1/3 of the cutting-edge radius. Therefore, in this study, the minimum chip thickness ranges from 0.27 to 0.37 μm. Thus, to observe

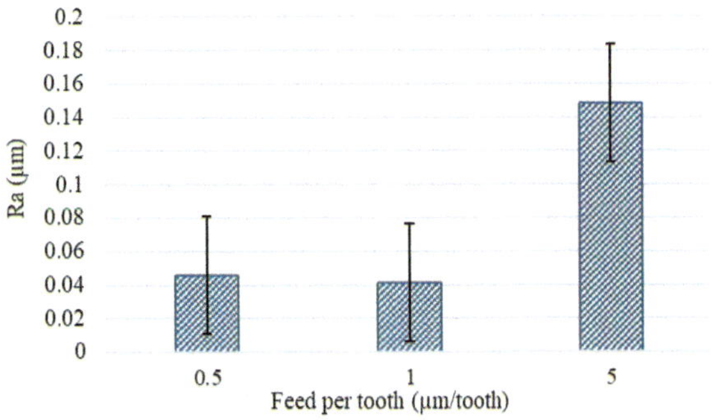

Fig. 3 Variation of surface roughness Ra with feed per tooth

the influence of the size effect on the quality of the machined surface, it would be necessary to evaluate feed per tooth values even smaller than 0.5 μm/tooth.

Figures 4, 5 and 6 show the images of the machined surfaces of the sample analysed for each feed per tooth.

In Fig. 4, it can be observed that for the lowest feed per tooth, the surface exhibits a better surface finish, with some pronounced feed marks (blue arrow), regions with material detachment (red circle) and areas of adhered material (black arrow). In Fig. 5, the feed marks appear regular, with small detachment regions (red circle) and some small particles of adhered material (black arrow), primarily at the entrance of the up-milling side. However, when the feed per tooth is increased to its highest value, the surface quality worsens considerably (Fig. 6), displaying noticeable material detachment (red circle), adhered material (black arrow) and the formation of internal burrs along the feed marks (red arrow). These internal burrs are formed due to

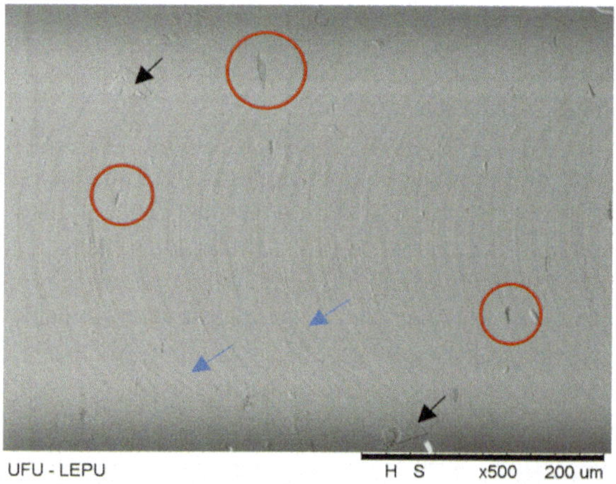

Fig. 4 Micromachined surface with feed of 0.5 μm/tooth

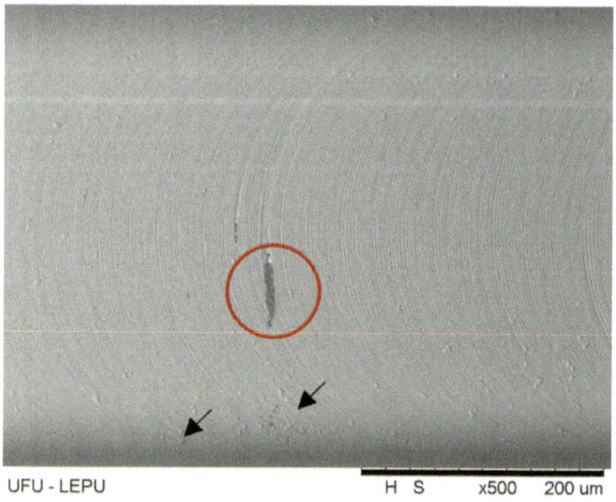

Fig. 5 Micromachined surface with feed of 1 μm/tooth

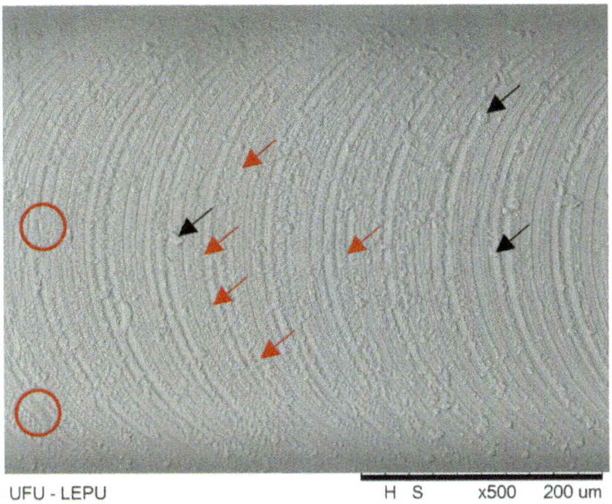

Fig. 6 Micromachined surface with feed of 5 μm/tooth

the occurrence of side flow, a phenomenon caused by high pressure exerted on the material left behind by the secondary cutting edge, leading to the plastic deformation of the material to the side [13].

4 Conclusions

When analysing the influence of the feed per tooth on the quality of the machined surface of each microchannel manufactured through micromilling with the parameters and material selected, it was possible to conclude that:

- No statistically significant difference was observed in the machined surface finish when using feeds of 0.5 and 1 μm/tooth, with average surface roughness (Ra) around 0.04 μm;
- Similar to the conventional milling process, the average surface roughness (Ra) increased with the feed per tooth, rising from 0.04 to 0.15 μm when the feed was significantly varied from 1 to 5 μm/tooth;
- From the tested parameters, the best feed rate in terms of surface roughness and surface morphology was 1 μm/tooth.

When observing the micromachined surfaces of each manufactured microchannel, it can be noticed that smaller feed per tooth results in a better surface finish, while larger feed per tooth causes material detachment, particle adhesion and formation of internal burrs due to side flow.

References

1. Câmara MA, Rubio JC, Abrão AM, Davim JP (2012) State of the art on micromilling of materials, a review. J Mater Sci Technol 28(8):673–685
2. O'Toole L, Kang CW, Fang FZ (2021) Precision micromilling process: state of the art. Adv Manuf 9:173–205
3. Attanasio A (2017) Tool run-out measurement in micromilling. Micromachines 8
4. Tamez-Tamez JI, Vázquez-Lepe E, Rodriguez CA et al (2023) Assessment of geometrical dimensions and puncture feasibility of microneedles manufactured by micromilling. Int J Adv Manuf Technol 126:4983–4996
5. Azo Materials, https://www.azom.com/article.aspx?ArticleID=2382. Last accessed 30 Jul 2024
6. Kajaria S, Chittipolu S, Adera S, Hung WN (2012) Micromilling in minimum quantity lubrication. Mach Sci Technol 16(4):524–546
7. Gomes MC, da Silva MB, Duarte MAV (2020) Experimental study of micromilling operation of stainless steel. Int J Adv Manuf Technol 111:3123–3139
8. Greco S, Kieren-Ehses S, Kirsch B et al (2021) Micromilling of additively manufactured AISI 316L: impact of the layerwise microstructure on the process results. Int J Adv Manuf Technol 112:361–373
9. Villares Metals, Aços Inoxidáveis, http://www.villaresmetals.com.br/villares/pt/Produtos/Acos-Inoxidaveis. Last accessed 25 Jul 2024
10. ISO 4288 (1996) Surface roughness
11. Machado AR, Silva MB, Coelho RT, Abrão AM (2015) Teoria da Usinagem dos Materiais, 2nd edn. Blucher, São Paulo
12. De Oliveira FB, Rodrigues AR, Coelho RT, De Souza AF (2015) Size effect and minimum chip thickness in micromilling. Int J Mach Tools Manuf 89:39–54
13. Shaw MC (2005) Metal cutting principles. Oxford University Press, New York

Feed Influence on Different Surface Roughness Parameters When Micro Milling Ti6Al4V

Aline Gonçalves dos Santos, Déborah de Oliveira, Milla Caroline Gomes, and Márcio Bacci da Silva

Abstract On micro machining, feed rate has great influence on the quality of the generated surface. Therefore, the present work aims to analyze the influence of the feed rate on the surface roughness in the Ti6Al4V alloy micro milling. For this, a 390 μm diameter micro tool, AlCrN coated, was used and tests were carried out with 19 different conditions, varying the feed rate from 0.05 to 5 μm/tooth, in steps of 0.05 μm/tooth, maintaining a constant cutting speed of 25 m/min, axial depth of cut of 40 μm, radial depth of cut of 390 μm. The roughness parameters Ra, Rq, Rz, Rt, Rsk and Rku were measured. The results showed that feed values from 0.20 to 0.30 μm/tooth provided better surface finishing, this feed corresponding to 20–30% of the value of the cutting-edge radius.

Keywords Micro milling · Ti6Al4V alloy · Surface roughness · Micro machining

1 Introduction

The need for small components, obtained through micro machining, is growing in several areas, but the medical area stands out, as smaller components can assist in less invasive treatments that cause less discomfort to patients. Among the biocompatible materials, titanium alloy Ti6Al4V stands out, due to its good mechanical properties and corrosion resistance. Mechanical micro machining is one way to manufacture

A. G. dos Santos (✉)
Faculty of Engineering, Federal University of Catalão, Catalão, GO, Brazil
e-mail: aline_santos@ufcat.edu.br

D. de Oliveira
Mechanical Engineering Department, University of Brasilia, Brasilia, DF, Brazil

M. C. Gomes
Department of Manufacturing Engineering and Materials, University of Campinas, Campinas, SP, Brazil

M. B. da Silva
School of Mechanical Engineering, Federal University of Uberlandia, Uberlandia, MG, Brazil

M. C. Gomes et al. (eds.), *Innovations in Brazilian Machining*, Lecture Notes in Mechanical Engineering, https://doi.org/10.1007/978-3-031-92819-2_12

these micro components. According to Câmara et al. [1], it stands out for achievement of different geometries with high quality and can be defined as a process in which cutting tools dimensions vary from 1 to 1000 μm.

Therefore, an adequate study of cutting parameters is essential for micro milling to avoid excessive plowing or even incorrect chip formation [2, 3]. Cutting conditions influence process performance, mainly wear and tool life and the integrity of the machined surface [4, 5]. Considering the medical application, the part must have high quality, however some problems can occur during micro milling, such as: low quality of the contours, burrs and defects on the surface, such as detachments and adhered material [6].

Those defects occur due to the micro cut characteristics, thereby, Brinksmeier et al. [7] suggest that studies be aimed at optimizing parameters to improve the finish of the part. Authors have suggested different approaches, Gomes et al. [8] indicated that the use of cutting fluid could reduce the presence of built-up-edges improving surface finish. Ziberov et al. [9], on the other hand, when micromilling Ti6Al4V, found that the use of coated tools could have great influence on the surface quality, also reducing the BUE formation.

When focusing on the burrs, some authors mentions that the presence of burrs not only affect the surface quality of micromilled surfaces burr is also associated with chip adherence and tool breakage [10]. Some authors like Silva et al. [11] suggested that varying the cutting velocity has low influence on the burr formation when micromilling Inconel, the authors however suggested the deburring as an alternative to improve the slot geometry.

In addition to this, recent studies shown that the micro cutting of Ti6Al4V alloy still needs investigation [12], to reduce burr formation, better comprehend the temperature distribution [13], tool wear and the details of the cutting mechanisms. Therefore, this work aims to evaluate the influence of feed rate on the surface quality of micro slots milled in Ti6Al4V.

2 Materials and Methods

To carry out the tests the machine tool used was a CNC micro milling machine with 3 linear axes Mini-mill/GX CNC machine from Minitech Machinery® was used, the axes have maximum offsets of 300, 228 and 228 mm in X, Y, and Z, directions, possessing a positioning resolution of 0.1 μm. Once the surface is the focus of this study, it is important to remark that the machine had reduced dimensions and is located in a granite inertial table.

The cutting tool selected was a flat-end micro mill with two cutting edges, coated with AlCrN, as suggested by the literature, and, a diameter of 390 μm was used. The tool edge radius has an approximate value of 1 μm, as shown in Fig. 1, the measurement was performed using scanning electron microscope (SEM-TM3000) images and Image J software.

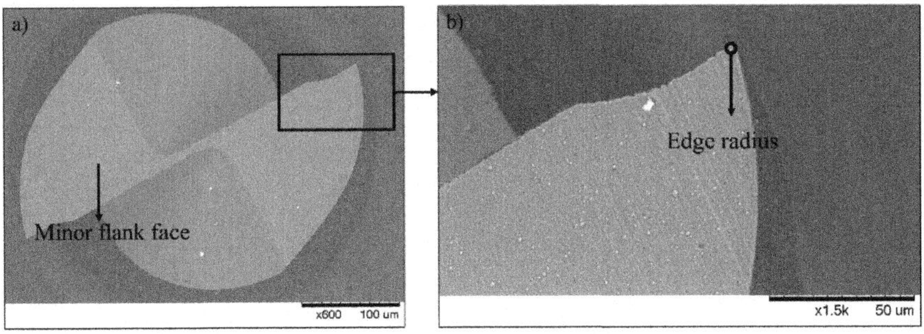

Fig. 1 Edge radius measurement of microtool, **a** top view of the microtool and **b** detail of the tool edge radius measurement

Table 1 Cutting parameters

Trial	1	2	3	4	5	6	7	8	9	10
Feed (μm/tooth)	0.05	0.10	0.15	0.20	0.25	0.30	0.35	0.40	0.45	0.50
Radial depth of cut	390 (μm)									
Axial depth of cut	40 (μm)									
Cutting velocity	25 (m/min)									

The tests were carried out dry, considering the medical application of the alloy, this condition was selected to avoid any contamination during the machining. The cutting parameters used were feed per tooth, varying from 0.05 to 5.0 μm/tooth, as can be observed on Table 1.

To measure the surface roughness of the slots, a Taylor Hobson Form Talysurf Intra 2 profilometer was used, the equipment possesses a resolution of 0.016 μm. To analyze the results, different surface roughness parameters were considered: Ra, Rq, Rz, Rt, Rsk and Rku. Three measurements were made on each slot analyzed and the results will be presented as the mean values. The sampling length (cut-off) used was 0.8 mm. To qualitative compare the surface quality, SEM images of the generated surfaces were also observed.

3 Results and Discussion

In Fig. 2, the graphs generated for the roughness parameters Ra, Rq, Rt and Rz, can be seen as a function of the feed values used. It should be noted that the standard deviation was not presented so as not to make the graph difficult to visualize.

For Ra and Rq, the minimum value was reached with a feed of 0.15 μm/tooth. However, the variation with feeds of 0.10–0.30 μm/tooth was small, with the greatest difference between values being less than 0.01 μm for Ra and 0.005 μm for Rq. From

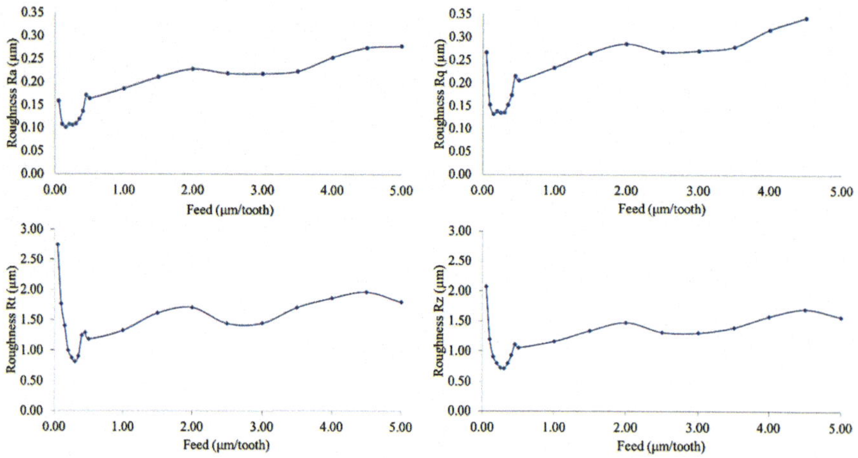

Fig. 2 Roughness of the slots when using different feeds (from 0.05 to 5 μm/tooth), axial depth of cut of 40 μm, radial depth of cut of 390 μm and cutting speed of 25 m/min

the feed rate of 0.30 μm/tooth to the value of 0.45 μm/tooth, the roughness increased rapidly, followed by a gradual increase until the last feed used. For Rt and Rz, the minimum value achieved was with a feed of 0.30 μm/tooth. However, the averages obtained were close for feeds of 0.20–0.35 μm/tooth. From this value (0.35 μm/tooth) until the feed of 0.45 μm/tooth, Rt and Rz also increased rapidly, followed by a gradual increase until the last feed used.

It is important to highlight that the Ra values obtained for the surface can be considered small and represent a good surface finish, in a review study evaluating different finishing methods the authors De Oliveira et al. [14] pointed out that in some cases the post processing is necessary not only to improve the surface by reducing the roughness but also to reduce the defects. To compare values with the presented on this work, in a surface enhanced with shot peening, for example, the Ra varied from 13.78 to 2.50 μm, even varying the post processing conditions [15]. Similarly, different processes like mechanical buffer, obtained Ra values all greater than 1.0 μm [14]; Ultrasonic cavitation abrasive finishing surfaces about 3.5 μm Ra [16] and only the abrasive processes being able to achieve values similar to the ones observed on this work [14], indicating that with the correct parameter selection the micromilling can enable the post processing to be avoided, reducing production costs, but also could achieve better results.

When studying the micro milling of Ti6Al4V alloy, Colpani et al. [17], applied two different cutting tools, with 800 μm diameter, coated with (Ti, Al)N, different helix angles and tool tips of 50 μm. The authors evaluated the influence of the tool wear on the surface roughness considering two different feeds: 10 and 20 μm/tooth. It was observed that the roughness values, as expected, were related to both the machining length, and consequently the tool tip radius and tool wear, and the feed. The Ra values varied between 0.1 and 0.3 μm. When comparing those results to the ones presented in this work, it is important to consider the theoretical roughness equation that consider that the maximum Ra will be a direct function of the feed per tooth (fz) and indirectly related to the tool tip edge radius (Er), as can be seen on Eq. 1 [17].

$$Ra \sim \frac{f_2^2}{8Er} \tag{1}$$

In this sense, on the work of, Colpani et al. [17], the values Ra for a feed that is 10 times higher than some of the used in this work, are similar not due to the influence of the feed, but the higher edge radius that their tool presented. This fact is important to highlight once micro tools with smaller diameters tend to present smaller edge radius, that could impact on the final surface, making it necessary to cautiously combine the cutting parameters. It can be reinforced that all the experiments on this work were performed with new tools with the same edge radius, to avoid variation on the edge radius and the wear influence. Therefore, the variation on the roughness parameters is from the feed variation.

As pointed out, the feed will have a great impact on the surface, focusing on it, it can be observed that the graphs present a similar behavior, at the beginning, the roughness tends to decrease with the increase in feed and, after a certain point, it starts to show a tendency to increase with this parameter, or it remains constants, as in the case of the Rsk and Rku parameters, as can be observed on Fig. 3. Thus, the graphs for Ra, Rq, Rz and Rt present two distinct regions. In one, the reduction in feed leads to a reduction in surface roughness, in the other, the reduction in feed causes its increase. As the feed rate increased, the roughness began to decrease until it reached a minimum value.

As for the Rsk parameter, it presented positive values with the three smallest feeds. With a feed of 0.2 μm/tooth, the Rsk became negative, and this value used provided a more regular surface in terms of the number of peaks and valleys, as it was the closest value to zero. From that point on, the values remained negative and did not undergo major variations. The Rsk parameter indicates the symmetry of the roughness profile. Negative Rsk values indicate a profile with a predominance of valleys. The Rku parameter evaluates the roundness of the roughness profile peaks. This parameter

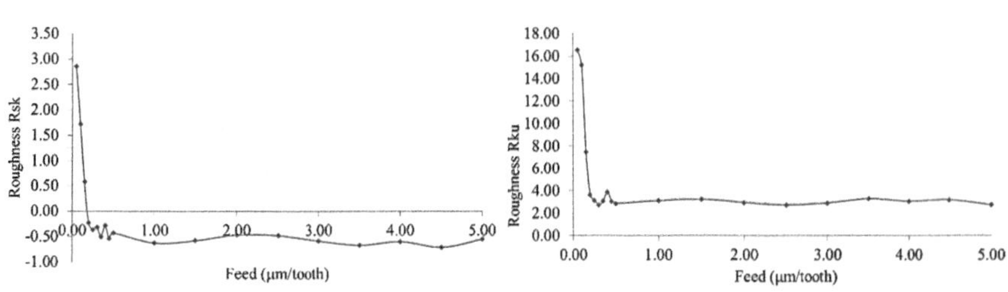

Fig. 3 Rsk and Rku roughness parameters of the slots when using different feeds (from 0.05 to 5 μm/tooth), axial depth of cut of 40 μm, radial depth of cut of 390 μm and cutting speed of 25 m/min

had similar behavior to Rsk. When using the three lowest feed values, this parameter presented high averages, greater than 3. From that point on, the roughness remained at values close to 3 μm, except for the slot in which the feed of 0.4 μm/m was used. Tooth, which presented a value of 3.886 μm.

When considering all parameters the range of feed values that provides better averages for roughness is between 0.20 and 0.30 μm/tooth, i.e., 20–30% of the value of the cutting-edge radius. Aslantas et al. [18] identified, in the micro milling of the Ti6Al4V alloy, that the feed of 0.5 μm/tooth provided the lowest value for roughness Ra, naming the region formed by values lower than this optimum point as the zone dominated by plowing. In the present work, when analyzing the graph obtained only for Ra, the feed rate that distinguishes the two regions with different behaviors is 0.15 μm/tooth. When reducing the values from this rate, the roughness increases. When increasing the feed value, the roughness also tends to increase. Therefore, the region encompassing values below this feed, according to the literature, is dominated by plowing. However, it is important to highlight that, when analyzing the graph for Rku, it is noted that the roughness was greater than 3 μm, using a feed of 0.15 μm/tooth. This indicates that the surface has many thin peaks and/or deep valleys. It was only after this feed rate that the Rku began to present values close to 3 μm, a surface finish similar to precision grinding [19]. Thus, suggesting the most appropriate parameters should be used according to the application of the part.

It is also important to analyze the surfaces generated with different parameters, as they can show different characteristics for different feed values used. In Fig. 4, the surfaces generated with three different feed rates can be seen. When analyzing the images of the slots obtained by SEM, the surface generated with the smallest feed (0.05 μm/tooth) presents many particles on the surface (black arrows) which may be an indication that the ploughing was dominant during machining. The occurrence of this phenomenon leaves marks due to the passage of the tool's cutting edge, which drags the material from the part without it being removed, increasing surface roughness.

As the feed increases, this effect decreases and the material is removed in the form of chips, thus, the surface roughness decreases, as can be seen on the surface generated with the feed of 0.20 mm/tooth. However, when increasing the feed value, from 1 μm/tooth, it was observed that feed marks became irregular. In Fig. 2, part of the surface generated with the feed of 4.5 μm/tooth can be observed, where indicative arrows exemplify an irregular mark present on the surface of the slot. Considering

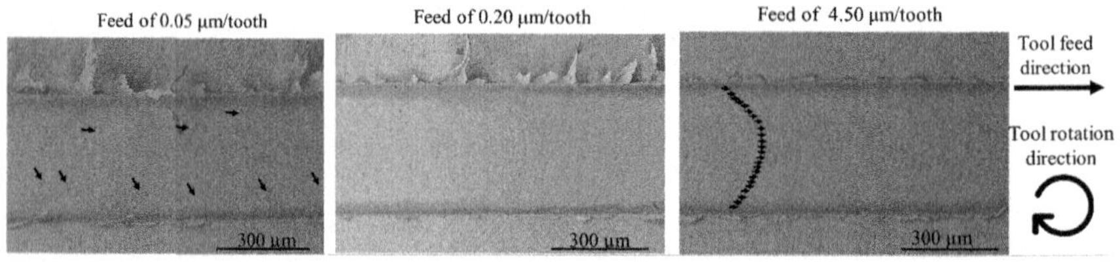

Fig. 4 Surfaces generated with feed rates of 0.05, 0.20 and 4.50 μm/tooth

that the marks became irregular with increasing feed rate, they could be caused by vibration of the machine tool. Thus, it is suggested that the use of feeds greater than 1 μm/tooth, under these analyzed conditions, causes vibration in the system, contributing to an increase in surface roughness.

4 Conclusions

After carrying out the tests and analyzing the results, it can be observed that, under the cutting conditions analyzed, the feed values of 0.20–0.30 μm/tooth provided lower surface roughness, when evaluating all roughness parameters. Therefore, in the present study, the cutting thickness that provides surfaces with lower roughness has values between 20 and 30% of the tool's edge radius. Finally, the work showed that for a more assertive definition of feed values, which provide lower surface roughness, it is important to evaluate different roughness parameters.

References

1. Camara MA, Rubio JCC, Abrão AM, Davim JP (2012) State of the art on micro-milling of materials, a review. J Mater Sci Technol 28(8):673–685
2. de Oliveira D, Gomes MC, Silva MB (2019) Spheroidal chip in micromilling. Wear 426–427:1672–1682
3. de Oliveira D, Gomes MC, de Oliveira GV, Santos AG, da Silva MB (2021) Experimental and computational contribution to chip geometry evaluation when micromilling Inconel 718. Wear 476:203658
4. Cheng K, Huo D (2013) Micro-cutting fundamentals and applications, 1st edn. Wiley, United Kingdom
5. Gomes MC, da Silva MB, Duarte MAV (2020) Experimental study of micromilling operation of stainless steel. Int J Adv Manuf Technol 111:3123–3139
6. Dornfeld D, Min S, Takeuchi Y (2006) Recent advances in mechanical micromachining. CIRP Ann 55:745–768
7. Brinksmeier E, Riemer O, Twardy S (2010) Tribological behavior of microstructured surfaces for microforming tools. Int J Mach Tools Manuf 50(4):425–430
8. Gomes MC, dos Santos AG, de Oliveira D, Figueiredo GV, Ribeiro KSB, De Los Rios GAB, da Silva MB, Coelho RT, Hung WNP (2022) Micro-machining of additively manufactured metals: a review. Int J Adv Manuf Technol 1–20
9. Ziberov M, de Oliveira D, da Silva MB, Hung WNP (2020) Wear of TiAlN and DLC coated microtools in micromilling of Ti-6Al-4V alloy. J Manuf Process 56:337–349
10. Wu X, Li L, He N (2017) Investigation on the burr formation mechanism in micro cutting. Precis Eng 47:191–196
11. Silva GP, da Silva MB, de Oliveira D (2023) Influence of abrasive deburring in indirect tool wear measurement in micromilling of Inconel 718. J Braz Soc Mech Sci Eng 45(5):262
12. Gonçalves MCC, Gomes MC, Stoeterau RL, Batalha GF, da Silva MB (2023) Influence of the material manufacturing process on micromilling Ti6Al4V alloy. Int J Adv Manuf Technol 129(1):23–35
13. Santos AG, da Silva MB, Oliveira GV, Jackson MJ (2023) Workpiece temperature investigation in micromilling of titanium alloy. CIRP J Manuf Sci Technol 47:193–204

14. de Oliveira D, Gomes MC, dos Santos AG, Ribeiro KSB, Vasques IJ, Coelho RT, da Silva MB, Hung NW (2023) Abrasive and non-conventional post-processing techniques to improve surface finish of additively manufactured metals: a review. Prog Addit Manuf 8(2):223–240
15. Calignano F, Manfredi D, Ambrosio EP (2013) Influence of process parameters on surface roughness of aluminum parts produced by DMLS. Int J Adv Manuf Technol 67:2743–2751
16. Tan K, Yeo SH (2017) Surface modification of additive manufactured components by ultrasonic cavitation abrasive finishing. Wear 378–379:90–95
17. Colpani A, Fiorentino A, Ceretti E, Attanasio A (2019) Tool wear analysis in micromilling of titanium alloy. Precis Eng 57:83–94
18. Aslantas K, Hopa HE, Percin M, Ucun I, Çicek A (2016) Cutting performance of nanocrystalline diamond (NCD) coating in micro-milling of Ti6Al4V alloy. Precis Eng 45:55–66
19. de Paiva RL, de Oliveira D, Ruzzi RS, Jackson MJ, Gelamo RV, Abrão AM, Da Silva RB (2023) Surface roughness and morphology evaluation of bearing steel after grinding with multilayer graphene platelets dispersed in different base fluids. Wear 523

Micromilling of AISI H13 Steel Manufactured by Directed Energy Deposition

Josenilton dos Santos Lopes, Marcio Bacci da Silva, Carlos José de Araujo, Marcus Vinícius de Oliveira, and Pietro Bedule Câmara

Abstract This study investigates the micromilling of AISI H13 manufactured by directed energy deposition, with a focus on the tool wear, surface integrity, and burr area formation. The parameters analyzed were the mean arithmetic deviation (Ra), the root mean square deviation (Rq), the skewness (Rsk), and the kurtosis (Rku). The microchannels measure 13.0 mm in length using micro end mills with a 400 μm diameter coated with (Al, Ti)N. The Vascomil MMA cutting fluid was applied using the minimum quantity lubrication technique. The input variables were a spindle speed of 10,000 rpm, a cutting speed of 12.57 m/min, a feed rate of 1.0 μm/tooth, a radial depth of cut of 400.0 μm, and an axial depth of cut of 40.0 μm. The wear was evaluated by comparing the worn profile of the used microtool on the secondary flank face with that of a new microtool. The results showed a roundness of the cutting-edges. Additionally, after micromachining 78 mm, the wear reached 11.91 μm (left cutting-edge) and 15.37 μm (right cutting-edge), corresponding to 3% and 4% of microtool diameter, respectively. The roughness Ra and Rq increased with the machined length. Moreover, the roughness parameter Rku decreased by 41.5%, indicating a transition from sharp shape peaks and valleys to rounded shapes.

Keywords H13 tool steel · Additive manufacturing · Micromilling · Tool wear · Roughness · Burr

J. dos Santos Lopes (✉) · M. B. da Silva · M. V. de Oliveira · P. B. Câmara
Faculty of Mechanical Engineering, Federal University of Uberlândia, Uberlândia, MG, Brazil
e-mail: josenilton.lopes@ufu.br

C. J. de Araujo
Department of Mechanical Engineering, Center of Science and Technology, Federal University of Campina Grande, Campina Grande, PB, Brazil

M. C. Gomes et al. (eds.), *Innovations in Brazilian Machining*, Lecture Notes in Mechanical Engineering, https://doi.org/10.1007/978-3-031-92819-2_13

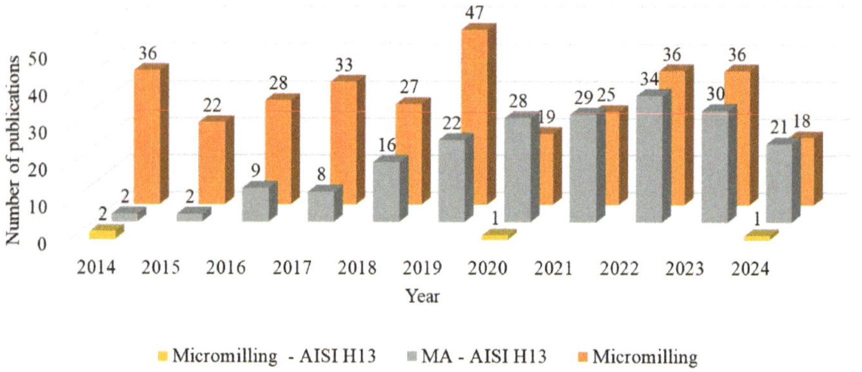

Fig. 1 Comparison of articles published on micromilling, AISI H13 steel manufactured by AM, micromilling of AISI H13 steel

1 Introduction

Micromilling can be used as a post-processing technique to mitigated inherent defects in additively manufactured (AM) products [1, 2]. O'Toole et al. [3] defined micromilling as a process restricted to microtool diameters smaller than 1000 μm. Brinksmeier and Preuss [4] related the definition material removal rate between 0.00001 and 1 $mm^3\ s^{-1}$. Moreover, the range of cutting parameters also contributes to the definition of micromiling [5–7].

Gomes et al. [7] compared the micromilling of AISI H13 tool steel manufactured by laser-beam directed energy deposition and hot rolling. In this work, the microtool wear did not affect surface roughness Ra. However, the authors did not investigate burr area formation. Recently, Lopes et al. [8] studied the micromilling of NiTinol under MQL conditions. The results showed that the burr area increased by 25% as the microtool wear increased.

The AISI H13 steel is used to manufacture die casting and forging dies due to its good wear resistance at elevated temperatures and fatigue resistance [9]. Recently, there has been an increase in interest in micromilling, as shown in Fig. 1. However, there is a lack of significant publications on micromilling of AISI H13 steel produced by additive manufacturing. In this context, this paper investigates tool wear, surface integrity, and burr formation on micromilling of AISI H13 steel fabricated by directed energy deposition.

2 Experimental Methodology

The AISI H13 steel manufactured by DED had an approximate hardness of 47 HRC, and width, length, and height of 13 mm, 12 mm, and 5 mm, respectively. Table 1 depicts the chemical composition obtained by energy dispersive spectroscopy (EDS).

Table 1 Chemical composition of the workpiece

C	Si	Mn	Cr	Mo	Al	Ni	V
7.552	0.049	0.419	6.559	0.483	0.105	0.278	1.332

The microtools manufactured by Mitsubishi Materials, featuring two flutes and a diameter of 0.4 mm, were employed in the experiments. The minor flank face (top view) is presented in Fig. 2.

The CNC Mini-Mill/GX machining center was utilized for the experiments (Fig. 3), with a maximum spindle speed of 60,000 rpm. The cutting fluid Vascomil MMS FA 2 was applied by the MQL technique with a pressure and flow rate of 0.23 MPa and 40.7 ml/h, respectively. The cutting parameters were selected based on the literature [7, 8], as shown in Table 2.

The wear was assessed by comparing the worn profile of the used microtool to that of a new microtool, analyzing the images of the secondary flank face. The illustration of wear measurement direction is depicted in Fig. 4, similar to those conducted by Gomes et al. [7].

The parameters analyzed were the mean arithmetic deviation (Ra), the root mean square deviation (Rq), the skewness (Rsk), and the kurtosis (Rku) using the Taylor Hobson profilometer with a resolution of 0.016 μm, using a cut-off and an evaluation length of 0.25 mm and 1.25 mm, respectively [10]. Additionally, the ImageJ software was utilized in burr area measurements, similar to those employed by Lopes et al. [8].

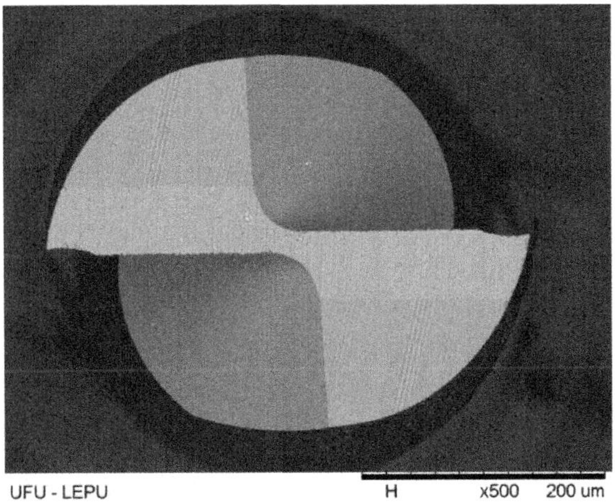

Fig. 2 Top view of the (Al, Ti)N coated microtool

Fig. 3 Experimental setup

Table 2 Cutting parameters

Cutting speed (m/min)	Feed rate (μm/tooth)	Axial depth of cut (μm)
12.57	1	40

3 Results

The wear evolution as a function of machining length is depicted in Fig. 5. The wear on the right cutting-edge reached 7.61 μm, after micromachining 13 mm. Thereafter, the wear rate diminished [8]. This behavior is similar to the wear behavior reported in the literature [8, 11, 12]. On the left cutting edge, wear reached 4.64 μm after the first microchannel and increased linearly. Additionally, the wear measured on the left cutting-edge was 11.91 μm and the wear measured on the right cutting-edge was 15.37 μm after machining 78 mm, corresponding to 3% and 4% of microtool diameter, respectively.

Figure 6 shows the change in microtool geometry on both cutting-edges after fabricating the last microchannel, Fig. 6b, evidencing the increase in microtool wear

Fig. 4 Wear measurements

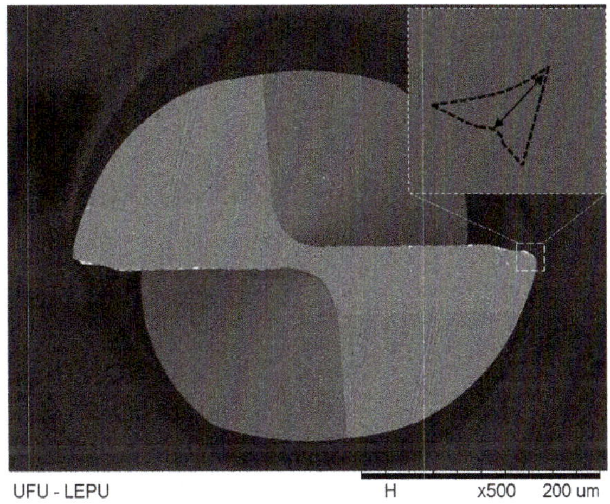

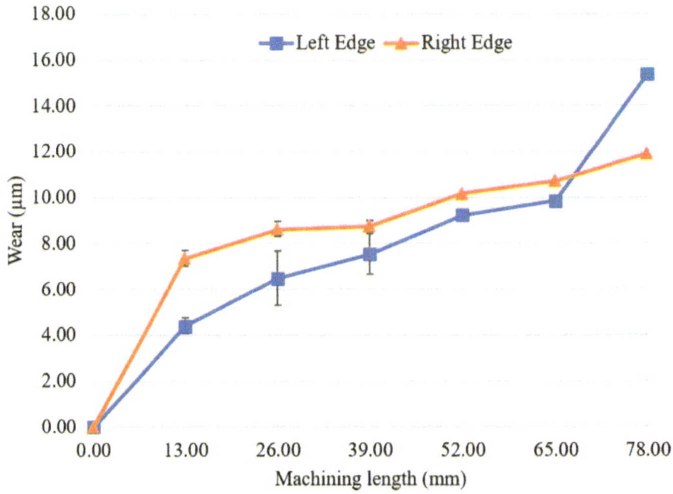

Fig. 5 Wear evolution

[13]. The increase in cutting-edge radius (r_e) is similar to that reported by Gomes et al. [7] and Lopes et al. [8]. Furthermore, a coating delamination was observed on the left cutting-edge, similar to reported by Aslantas et al. [13].

Figure 7 shows the average values of the mean arithmetic deviation (Ra), the root mean square deviation (Rq), the skewness (Rsk), and the kurtosis (Rku).

In the last microchannel (78 mm), the roughness value Ra and Rq was found to be (0.051 ± 0.002) μm and (0.062 ± 0.003) μm, respectively (Fig. 7a, b). These values represent an increase of 82% and 63%, respectively. Additionally, this roughness behavior differs from that reported by Gomes et al. [7] and Lopes et al. [8].

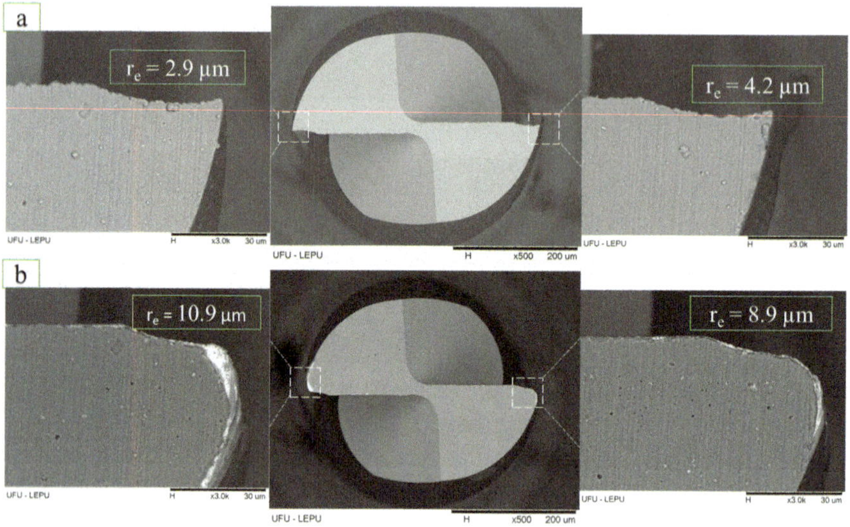

Fig. 6 Microtool after machining length of **a** 0 mm and **b** 78 mm

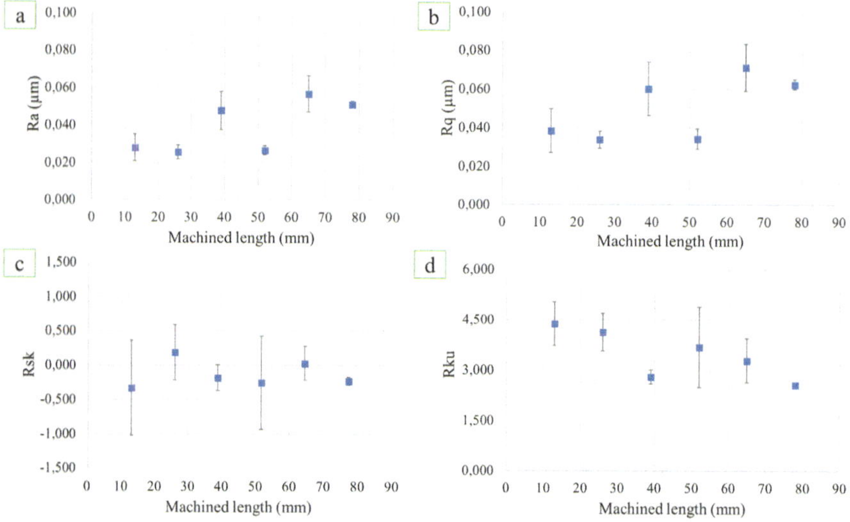

Fig. 7 **a** Ra, **b** Rq, **c** Rsk and **d** Rku obtained for each microchannel

Besides, the roughness Ra obtained is significantly lower than those reported by Bonaiti et al. [14] when micromilling of titanium alloy Ti-6Al-4 V manufactured by DED, Brito et al. [15] when investigated the micromilling of titanium alloy Ti-6Al-4V manufactured conventionally, Aramcharoen and Mativenga [6] when micromiling

AISI H13 (45 HRC), and Tansukatanon et al. [16] when micromilling AISI 304 stainless steel.

Further, in Fig. 7c, in the third microchannel (39 mm) and last microchannel (78 mm), the roughness value Rsk was (−0.189 ± 0.183) μm and (−0.231 ± 0.055) μm, respectively. This result suggests that valleys are predominant in these microchannels, similar to those highlighted by Silva [17].

In Fig. 7d, after micromachining 78 mm, the roughness Rku values diminish by 41.5% compared to the first microchannel (13 mm). This result suggests a transition from peaks and valleys with sharp shapes to rounded shapes [17]. The roughness Rku behavior differs from that reported by Silva [17] when micromilling 316L stainless steel manufactured by selective laser melting (SLM).

The feed marks and the microgrooves on the machined surface are presented in Fig. 8. The width of these marks was approximately (1.1 ± 0.2) μm, which is similar to the feed rate of 1.0 μm/tooth.

Figure 9 shows the burr area tends to increase with the machined length, Fig. 9, similar to those highlighted by Lopes et al. [8].

The highest burr area on the down-milling side (101,879.98 μm^2) is approximately 92% higher than the burr area on the up-milling side (52,973.20 μm^2), after a machining length of 78 mm, as can be seen in Fig. 10a, b. The microtool wear and size effect explains this result [8, 18].

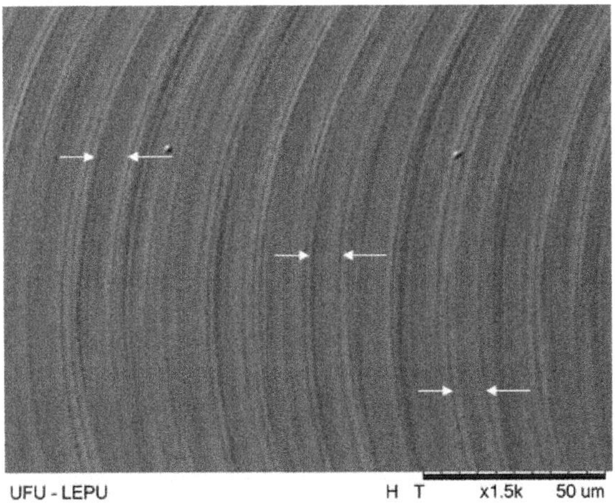

Fig. 8 Machined surface after machining length of 78 mm

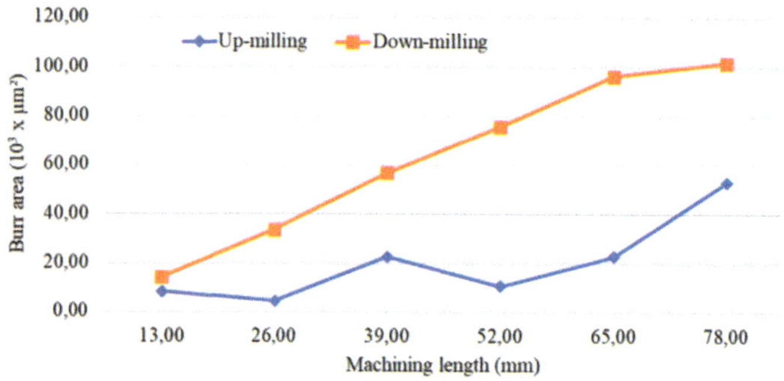

Fig. 9 Burrs area behavior

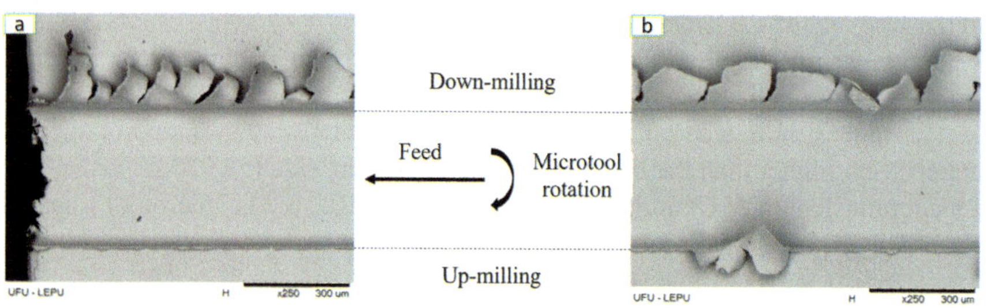

Fig. 10 Microchannels after machining length of **a** 26 mm and **b** 52 mm

4 Conclusions

The roundness of the cutting-edge contributed to the increase in surface roughness (Ra and Rq) and burr area. The roughness Rku values decreased by 41.5% with the increase in the machined length. The highest burr area on the down-milling side was 92% higher than the up-milling side.

Acknowledgements The authors would like to thank: CAPES, CNPq, FAPEMIG, and the Graduate Program in Mechanical Engineering at the Federal University of Uberlândia for financial support. We also thank the organizing committee of the XXVI Machining Colloquium.

References

1. Gomes MC, Santos AGD, De Oliveira D, Figueiredo GV, Ribeiro KSB, De Los Rios GaB, Da Silva MB, Coelho RT, Hung WNP (2021) Micro-machining of additively manufactured metals: a review. Int J Adv Manuf Technol 118:2059–2078
2. Azarniya A, Colera XG, Mirzaali MJ, Sovizi S, Bartolomeu F, Marek SW, Wits WW, Yap CY, Ahn J, Miranda G, Silva FS, Hosseini HRM, Ramakrishna S, Zadpoor AA (2019)

Additive manufacturing of Ti–6Al–4V parts through laser metal deposition (LMD): process, microstructure, and mechanical properties. J Alloy Compd 804:163–191
3. O'Toole T, Kang WC, Fang ZF (2021) Precision micro-milling process: state of the art. Adv Manuf 9:173–205
4. Brinksmeier E, Preuss W (2012) Micro-machining. Philos Trans Roy Soc Math Phys Eng Sci 370(1973):3973–3992
5. Rodrigues A, Jasinevicius R (2017) Machining scale, microfabrication and precision engineering [s.l.]. Elsevier, pp 27–68
6. Aramcharoen A, Mativenga PT (2009) Size effect and tool geometry in micromilling of tool steel. Precis Eng 33(4):402–407
7. Gomes MC, da Silva MB, Hung WNP, Coelho RT (2023) Experimental study of micro-milling of H13 tool steel manufactured by directed energy deposition. COBEF 2023. Lecture notes in mechanical engineering. Springer
8. Lopes JS, Câmara PB, Davis R, Araujo CJ, da Silva MB, Machado AR (2025) Insights into the fabrication of microchannels on biomedical NiTinol. CIRP J Manuf Sci Technol 56:106–118
9. Bidare P, Jiménez A, Abdelkhalek SM, Khaled I, Hassan A, Abdallah R, Essa K (2023) Hybrid additive manufacturing of AISI H13 through InterLayer machining and laser re-melting. Mater Today Commun 37
10. ISO 4288:1996 (1996) Geometrical product specifications (GPS)—Surface texture: profile method—rules and procedures for the assessment of surface texture
11. Machado Á, Abrão A, Coelho R, Silva M (2015) Teoria Da Usinagem Dos Materiais, 3rd edn. Blucher
12. Ziberov M, da Silva MB, Jackson M, Hung WNP (2016) Effect of cutting fluid on micromilling of Ti-6Al-4V titanium alloy. Procedia Manuf 5:332–347
13. Aslantas K, Hascelik A, Çiçek A (2022) Performance evaluation of DLC and NCD coatings in micro-milling of Al7075-T6 alloy. J Manuf Process 81:976–990
14. Bonaiti G, Parenti P, Annoni M, Kapoor S (2017) Micro-milling machinability of DED additive titanium Ti-6Al-4V. Procedia Manuf 10:497–509
15. Brito LC, Gomes MC, De Oliveira D, Da Silva MB, Duarte MAV (2023) Vibration features for indirect monitoring of end micromilling process. Precis Eng 79:7–15
16. Tansukatanon S, Tangwarodomnukun V, Dumkum C, Kruytong P, Plaichum N, Charee W (2019) Micromachining of stainless steel using TiAlN-coated tungsten carbide end mill. Procedia Manuf 30:419–426
17. Silva L (2022) 316L Stainless steel micromilling made by selective laser melting (SLM) using carbide tools with different types of coatings. Ph.D. Thesis, Federal University of Uberlandia, Uberlandia-MG, p 155
18. dos Santos AJ, Reis BCM, Pereira NFS (2023) Tribological effects of micromilling of hardened AISI D2 steel on tool wear and top burr formation. Int J Adv Manuf Technol 127:5327–5341

Micromilling of H13 Manufactured by DED by $NaNO_3$ Coated Microtool

Lazara Eduarda da Silva Damacena, Josenilton dos Santos Lopes, Milla Caroline Gomes, and Marcio Bacci da Silva

Abstract The wear analyses in microtools remains challenging due to the impossibility of using technical standards applied in macromilling. Moreover, surface finish analysis is crucial when micromachining parts manufactured by additive manufacturing, as this process enhances surface finish and dimensional accuracy. Therefore, this study aimed to analyze the wear and surface roughness in microchannels fabricated through micromilling on AISI H13 tool steel, using a microtool with a diameter of 0.4 mm coated with $NaNO_3$. The tests were performed on CNC milling system with a maximum spindle speed of 60,000 rpm. Minimum Quantity Lubrication (MQL) was employed for cutting fluid application. It uses a cutting speed of 12.6 m/min, feed rate of 5 μm/tooth, radial depth of cut of 400 μm and axial depth of cut of 40 μm. The results showed that the Ra roughness values machining with the $NaNO_3$-coated tool ranged from 0.111 to 0.278 μm. After machining the last channel (130 mm), the wear of the $NaNO_3$-coated tool reached (10.45 ± 2.71) μm. Additionally, no statistical differences were found between the Ra values of the first machined microchannel and the last microchannel. In future work, new experiments with statistical analyses will be conducted to enable complementary analyses.

Keywords Micromachining · Micromilling · Wear · Surface finish · H13 tool steel

1 Introduction

The miniaturized components are used in the biomedical, automotive, and aerospace industries. These applications demand extremely low dimensional tolerances, which are achieved through the micromachining process [1]. Micromachining is the manufacturing process used to achieve surface quality and higher dimensional accuracy

L. E. da Silva Damacena · J. dos Santos Lopes (✉) · M. B. da Silva
Faculty of Mechanical Engineering, Federal University of Uberlândia, Uberlândia, MG, Brazil
e-mail: josenilton.lopes@ufu.br

M. C. Gomes
Department of Manufacturing Engineering and Materials, University of Campinas, Campinas, SP, Brazil

M. C. Gomes et al. (eds.), *Innovations in Brazilian Machining*, Lecture Notes in Mechanical Engineering, https://doi.org/10.1007/978-3-031-92819-2_14

in the final products. One of the many definitions for micromachining is based on the cutting tool diameter, which needs to be between 1 and 1000 μm [1, 2]. Another definition is when the cutting thickness is between nanometers and hundreds of micrometers [3].

As components manufactured by additive manufacturing have low dimensional accuracy and surface finish, micromachining processes have been used as a post-processing technique for these parts [3]. Even with the undesirable characteristics mentioned, additive manufacturing has been increasingly adopted due to the ability to produce parts with complex geometries and, in addition to rebuilding and restoring parts that have suffered wear or been damaged, it is possible to create metallic elements from a base [4, 5]. One of the additive manufacturing methods, which stands out, is direct energy deposition (DED). In this technique, a source of thermal energy is used, whose heat is generated through a laser beam, electrons or electric arc, to fuse and join materials. The materials are typically in the form of filament or metal pow-der. The material is melted and deposited on the substrate when it comes into contact with the heat source [6].

Although micromachining is a form of machining on a reduced scale, it has some differences from conventional machining. This is caused by the miniaturization of the workpieces, cutting tools, and overall process [7]. In micromachining, the wear of the cutting tool alters its geometry and is considerably impacted by the temperature variations present during the process, which is gradual and continuous [8]. In addition, wear affects the quality and costs of machining processes [9]. In micromachining, these consequences are more critical due to the micrometric dimensions involved in the process. Therefore, it is necessary to reduce wear in micromachining and one of the ways to reduce it, and consequently increase the cutting tool life, is to apply coatings on the surface of the cutting tool [10].

Therefore, the objective of this work is to analyze the wear of a carbide tool coated with $NaNO_3$ and surface roughness during the micromilling of H13 tool steel, manufactured by additive manufacturing by the process called directed energy deposition (DED).

2 Experimental Methodology

The material used in the experiments was AISI H13 tool steel, manufactured by the additive manufacturing process called directed energy deposition (DED). This material is used in the manufacture of molds and dies. The sample has a hardness of 513 HV, and its chemical composition by weight percentage is shown in Table 1. The composition was obtained by EDS using the Hitachi HighTechnologies® TM3000 SEM. It is noted that the large amount of carbon is due to the inaccuracy of the SEM system in detecting this element.

The machine tool was a Mini-mill/GX CNC micromilling machine, manufactured by Minitech Machinery Corporation®. This machine has a maximum spindle rotation of 60,000 rpm, 3 axes with a positioning resolution of 0.1 μm each.

Table 1 Chemical composition in the percentage of the weight of the AISI H13 sample

	Chemical composition (% wt)							
C	Si	Mn	P	Cr	Mo	Al	Cu	V
8.155	1.128	0.431	0.081	5.286	1.049	0.019	0.004	0.745

The microtool used was made of carbide, coated with $NaNO_3$, manufactured by Performance tools, with two cutting edges, a cutting diameter of 400 μm, a cutting length of 800 μm, total length of 40 mm and shank length of 4 mm. The machine tool is installed on an inertial table, specially designed to reduce the influence of vibrations from external sources during micromachining.

The experiments were conducted using Blaser's Vascomill MMS FA2 cutting fluid, applied by the minimum quantity lubrication (MQL) technique, with a flow rate of 130 ml/h and a pressure of 6 bar. The fluid nozzle was positioned approximately 10 mm from the microtool and inclined at 20°, as shown in Fig. 1. The micromilling parameters are presented in Table 2. In total, 20 microchannels were manufactured.

After machining the microchannels, the microtools were examined in the SEM for analysis. The wear of the microtool was evaluated by measuring the reduction of the diameter on the clearance surface, as shown in Fig. 2. To minimize measurement errors, the secondary flank face images (top images) of the worn micro mills were overlaid on the images of the new tools using Adobe Photoshop CC software. The diameter of each microtool was then measured using ImageJ software, following the methodology described by Gomes [3].

The arithmetic mean deviation (Ra) was measured using a Taylor Hobson® profilometer, model Form Talysurf Intra, which features a resolution of 16 nm and

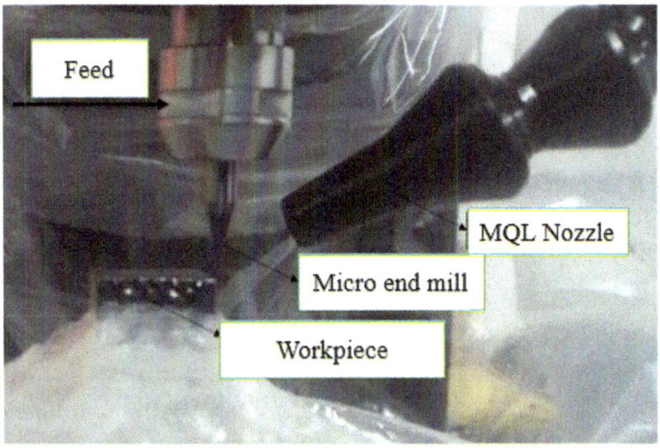

Fig. 1 Experimental configuration

Table 2 Micromilling parameters

Cutting speed vc (m/min)	Feed rate fz (μm/tooth)	Table feed vf (mm/min)	Depth of cut ap (μm)
12.6	5	100	40

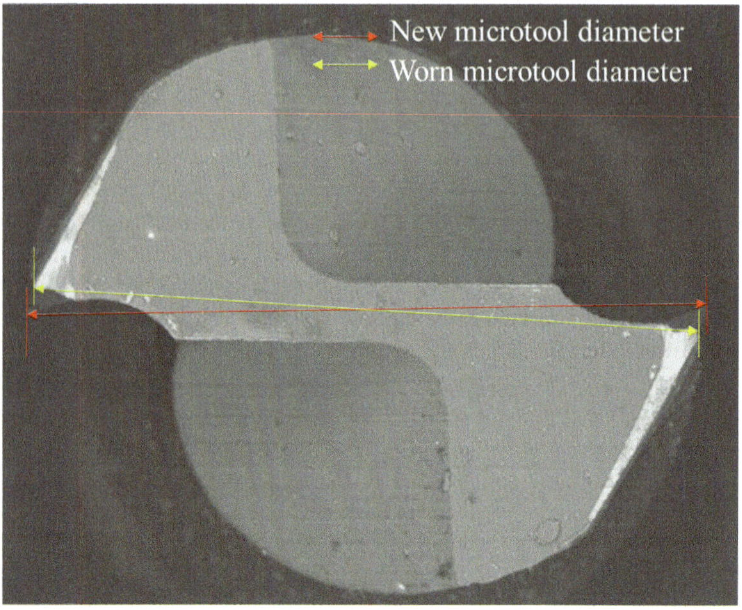

Fig. 2 Wear measurements

a diamond-tipped stylus with a radius of 2 μm. Measurements were conducted in a temperature-controlled environment maintained at approximately 20 ± 2 °C. Three measurements perpendicular to the feed marks were taken from different regions of each microchannel, using a Gaussian filter, a cut-off of 0.8 mm and an evaluation length of 4 mm, according to ISO 4288 [11] (Fig. 3).

3 Results and Discussion

After machining the last channel (130 mm), the wear of the $NaNO_3$-coated tool reached 10.45 ± 2.71 μm, corresponding to 2.6% of the diameter of the microtool. This result was similar to that Alonso [12] reported when micromilling of AISI H13 tool steel using (Al, Ti) N-coated microtool. Figure 4 shows the cutting-edges of the microtool coated with $NaNO_3$ before and after ten microchannels, it is possible to observe there the change in the geometry of the microtool, evidenced by chipping and loss of coating cohesion with both the cemented carbide substrate and the $NaNO_3$ coating.

Figure 5 shows the mean arithmetic deviation Ra behavior for each microchannel manufactured, with a standard deviation with a confidence interval of 68.27%.

As can be seen in Fig. 5, the roughness Ra values ranged from 0.111 to 0.278 μm. Gomes et al. [13] observed similar roughness Ra values in the micromilling of

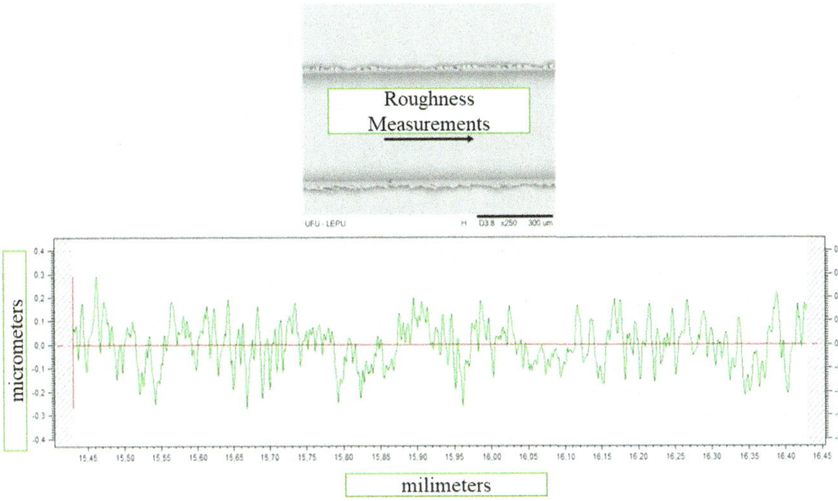

Fig. 3 Measurement of the roughness in the microchannel and roughness profile obtained

conventionally manufactured 316L stainless steel. In addition, this value range is lower than the roughness Ra range determined by Diniz, Marcondes and Copinni [14] for the grinding processes, which varies from 0.20 to 1.60 μm. This shows that in micromilling of AISI H13 tool steel manufactured by DED, the surface finish can be better than those resulted by the grinding process, which is a finishing process.

Regarding the behavior of the roughness values Ra, Fig. 5, it is possible to observe there are no statistical differences between the values obtained for the first channel (13 mm) and the last channel (130 mm). This result differs from those obtained by Gomes et al. [15] when comparing the micromilling of AISI H13 manufactured by DED (513 HV) and hot rolling (207 HV), using (Al, Ti) N-coated carbide. The authors reported a tendency of reduction with the increase of the machining length.

Figure 6 shows the images of the machined surfaces of the first microchannel and last microchannel machined with a $NaNO_3$-coated tool. It is noted that the surface on both channels has evident feed marks of the microtool.

4 Conclusions

The following conclusions can be drawn from this study:

Tool wear is evidenced by chipping and the loss of cohesion between the coating and both the cemented carbide substrate and the $NaNO_3$ coating.

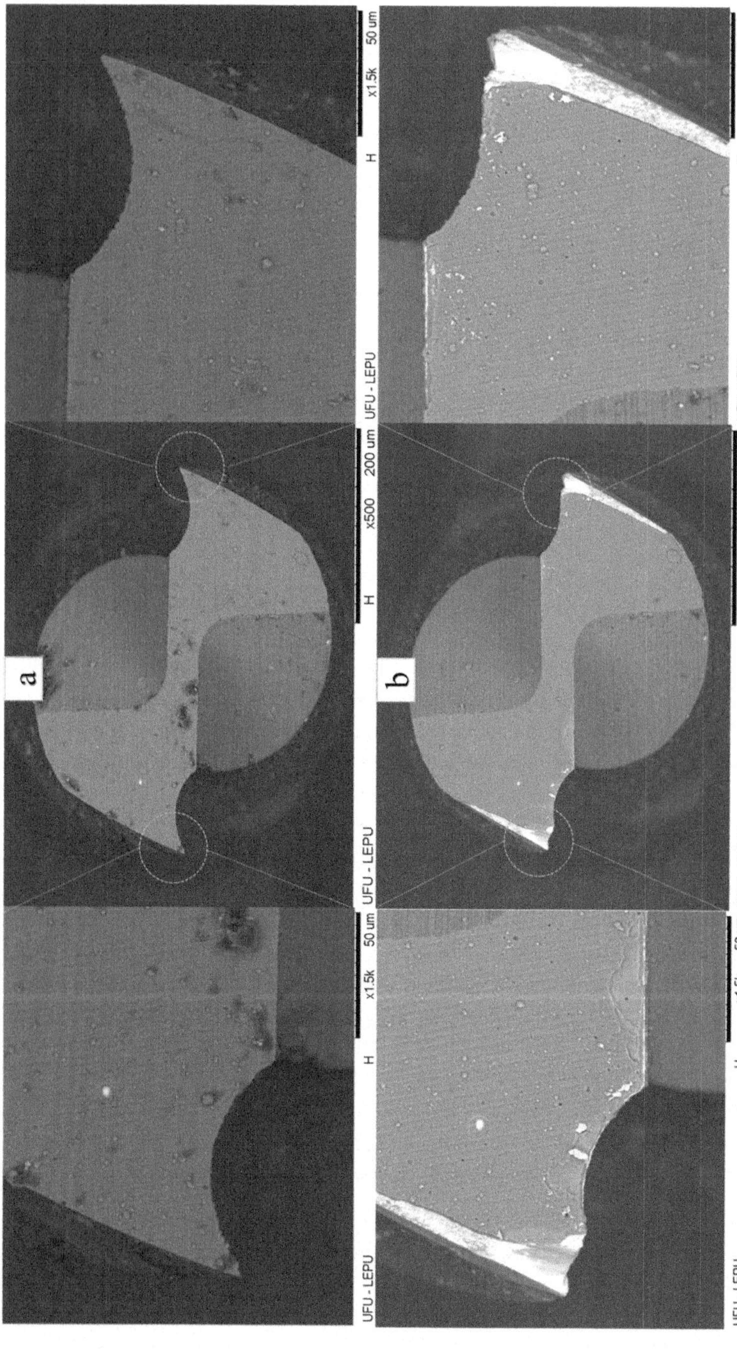

Fig. 4 Microtool after machining length of **a** 0 mm and **b** 130 mm

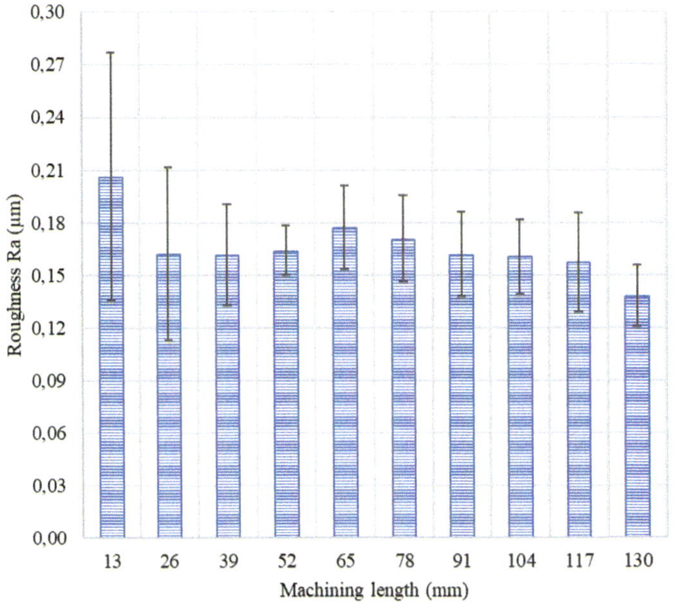

Fig. 5 The roughness Ra as a function of machined length

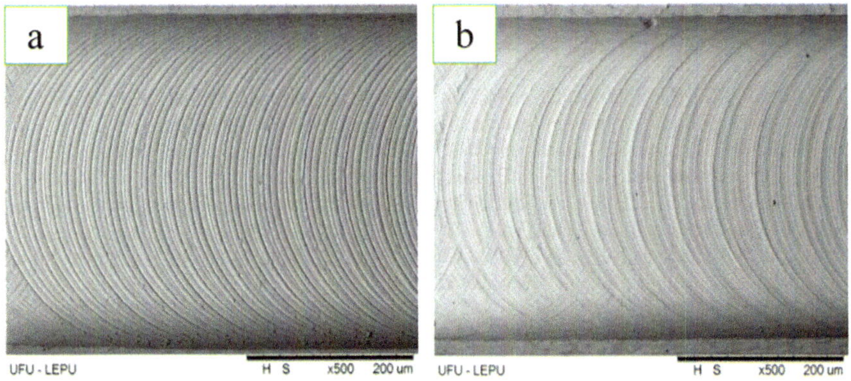

Fig. 6 Machined surface of **a** first microchannel and **b** last microchannel of the sample machined with $NaNO_3$ coated tool

When analyzing the surface roughness Ra of the microchannels manufactured in the ABNT H13 tool steel manufactured by DED, no statistical differences between the values were obtained, when comparing the roughness of the first machined micro-channels with the roughness of the last microchannel manufactured.

Further investigations will be observed, such as wear evolution after fabrication of each microchannel and statistical analysis to complement the observations already made.

References

1. Masuzawa T (2000) State of the art of micromachining. CIRP Ann Manuf Technol 49:473–488
2. Cheng K, Huo D (2013) Micro-cutting—fundamentals and applications. Wiley, United Kingdom, p 366
3. Gomes MC (2022) Micromilling of AISI 316L stainless steel manufactured by casting and laser-directed energy deposition. Doctoral Thesis, Federal University of Uberlândia, Uberlândia
4. Zhai Y, Lados DA, Lagoy JL (2014) Additive manufacturing: making imagination the major limitation. JOM 66(5):808–816
5. Liu R, Eaton E, Yu M, Kuang J (2017) An investigation of side flow during chip formation in orthogonal cutting. Procedia Manuf 10:568–657
6. ISO 52900 (2018) Additive manufacturing—general principles—terminology, 1st edn, pp 1–25
7. Berestovskyi DV (2013) Surface finish modeling in micromilling of biocompatible materials. Master of Science, Texas A&M University
8. Machado AR, Abram AM, Coelho RT, da Silva MB (2015) Theory of machining of materials, 3rd ed. Edgard Blucher Publishing, São Paulo, SP
9. Uhlmann E, König J, Sammler F, Richarz S (2013) Tribology of treated and coated cutting tool surfaces. Encycl Tribol 3896–3905
10. Chen N, Li HN, Wu J, Li Z, Li L, Liu G, He N (2021) Advances in micro milling: from tool fabrication to process outcomes. Int J Mach Tools Manuf 160:103670
11. ISO 4288:1996 (1996) Geometrical product specifications (GPS)—surface texture: profile method—rules and procedures for the assessment of surface texture
12. Alonso T (2023) Experimental analysis of the effect of cutting conditions on micromilling of tool steel AISI H13. Master of Science, Federal University of Uberlândia
13. Gomes M, Da Silva MB, Duarte MAV (2020) Experimental study of micromilling operation of stainless steel. Int J Adv Manuf Technol 111(11):3123–3139
14. Malkin S, Guo C (2008) Grinding technology—theory and applications of machining with abrasives. Industrial Press, New York
15. Gomes MC, da Silva MB, Hung WNP, Coelho RT (2023) Experimental study of micro-milling of H13 tool steel manufactured by directed energy deposition. COBEF 2023. Lecture Notes in Mechanical Engineering. Springer

Tool Wear in Micro End Milling of Ultrafine Grain Steels

Flávio Henrique Manarelli, Izabel Fernanda Machado, and Alessandro Roger Rodrigues

Abstract Micro-machining is one of the processes feasible for generating geometries or parts with precision and complexity regarding feature dimensions smaller than 1 mm. However, the achievement of enhanced machining performance and product quality requires a prior understanding on the tool-workpiece interface. This research determined the influence of cutting strategy and feed per tooth on tool wear and specific cutting force when cutting the ultrafine grain steel COS AR60 under end milling condition. Machining tests were performed in a CNC machining center Romi D600 provided with a high-speed spindle (60,000 rpm) without cutting fluid application. TiAlN coated carbide end mill (Ø 800 μm) was used in linear tool path with 60 m/min cutting speed, 160 μm depth of cut and feed per tooth of 3 and 10 μm/z. Analysis of Variance (ANOVA) with confidence interval of 95% was applied by considering one run and two replications. The results showed that tool wear levels and damages were influenced by tool feed per tooth and cutting strategy. Additionally, chipping damage during face milling with higher feed per tooth was predominant. Adhesion and abrasion wear mechanisms were significant for the same cutting pass under slot milling and down milling at a lower feed per tooth. Down milling and slot milling with 3 μm/z feed per tooth leaded to a better performance in terms of tool wear level on the major flank face. Specific cutting force decreased with feed per tooth and increased when milling channel.

Keywords Micromilling · Tool wear · Ploughing · Specific cutting force

F. H. Manarelli (✉)
Federal Institute of Education, Science and Technology, São Paulo, SP, Brazil
e-mail: flaviomanarelli@usp.br; flavio.manarelli@ifsp.edu.br

F. H. Manarelli · I. F. Machado
University of São Paulo, São Paulo, SP, Brazil

A. R. Rodrigues
University of São Paulo, São Carlos, SP, Brazil

M. C. Gomes et al. (eds.), *Innovations in Brazilian Machining*, Lecture Notes in Mechanical Engineering, https://doi.org/10.1007/978-3-031-92819-2_15

1 Introduction

The level of detail, complexity, and surface finish of downsized parts has been steadily increasing. Along with these requirements many parts have high demands in the medical field and embedded systems within the automotive and aerospace sectors. Therefore these components often need to be produced with special alloys, resulting in additional manufacturing challenge [1]. In this context, micromilling is a highly suitable process for producing miniaturized parts from various materials. Combined with its high versatility and dynamic movements micromilling can create a wide range of geometric profiles in the produced parts, thereby meeting the increasingly specific geometric and dimensional demands of the current market [2].

In the case of micromachining its principles are quite similar to conventional machining, such as the kinematics associated with the relative movements between the tool and the workpiece. However, as the scale of the process decreases, phenomenological differences arise that prevent direct correlations with conventional machining processes. In micromachining, it is considered that the chip is formed with each pass of the tool teeth over the generated surface of the workpiece, and the tool is assumed to be perfectly sharp since the cutting edge radius is much smaller than the feed per tooth and the cutting thickness, becoming its influence insignificant in the process [3]. These assumptions are made considering that the tool cutting edge radius used in conventional cutting has dimensions on the order of micrometers, compared to the cutting thickness (h) and the typically used tool feeds, which range from millimeters to tenths of millimeters. Thus, it is observed that the chip formation process predominantly occurs by shearing [4].

Conversely in microcutting the tools have diameters ranging from 25 to 1000 μm, with depths of cut and tool feeds on the order of micrometers [5], and the cutting thickness varies from 0.1 to 200 μm [6]. Additionally, the tool cutting edge radius can be comparable to or even larger than the cutting thickness. In such cases, material removal by shearing is impaired given the highly negative rake angle [4]. Therefore it is crucial to consider a minimum cutting thickness that is sufficient to ensure effective material removal since microscale cutting processes exhibit specific characteristics related to the named size effect [7]. Cheng and Huo [8] describe that this phenomenon occurs when the cutting thickness is comparable to or even smaller than the tool cutting edge radius or the workpiece grain size.

The minimum cutting thickness (h_m) is defined as a threshold value sufficient to yield chip. Various studies aim to quantify this value with results ranging from 20 to 40% of the tool cutting edge radius (r_e), a variation attributed to the strong dependence on the machined material [8]. Thus, when the cutting thickness is less than this minimum value, the part material is subjected to elastoplastic deformation process without effective chip removal by causing the ploughing phenomenon [9].

In micromilling the variation in cutting thickness due to the tool rotation can start and/or end with a value smaller than the minimum cutting thickness. Regarding the cutting thickness and minimum thickness, Chae et al. [10] observed and characterized three distinct situations for the interaction between the micro end mill and the workpiece during machining:

- $h < h_m$: The ploughing phenomenon is predominant and the workpiece material volume in front of the tool cutting edge radius is elastoplastically deformed followed by elastic recovery;
- $h \approx h_m$: Not every the workpiece volume to be removed is transformed into chips i.e., a portion is removed by shearing while another one is deformed due to the ploughing phenomenon;
- $h > h_m$: Material removal occurs primarily by shearing and the chips are fully formed.

Therefore, it is worthy to apply cutting parameters that ensure the cutting thickness is predominantly greater than a minimum value. Oliveira et al. [9] highlight that if the cutting thickness equals a minimum, chip formation will occur but without maintaining the workpiece surface quality due to the deformations caused by the ploughing effect. Furthermore, in the presence of the ploughing effect acting at the tool edge and at the tool-workpiece interface, the specific cutting force will increase exponentially [11]. This growth is inversely proportional to the cutting thickness and is associated with the ploughing effect.

According to Campos et al. [4], when the ploughing phenomenon is present tool wear and cutting forces can increase fast. Furthermore, with larger tool wear the tool cutting edge radius will change by increasing the friction and intensifying the ploughing phenomenon, thereby compromising surface finish and production costs [3].

As aforementioned, tool wear increases the cutting edge radius given the material loss on both the tool rake and clearance surfaces. Thus, the wear mechanisms need to be better understood and characterized at microcutting. Additionally, the combination of various wear mechanisms commonly takes place during the cutting process. However, observing and monitoring these mechanisms are not an easy task especially at reduced scales. Nonetheless, detailed studies are essential since significant tool wear evolution can reach undesirable magnitudes, reduces tool life, or even lead to tool breakage by impacting the workpiece surface quality [2].

Colpani et al. [12] mention that ISO 8688 standard [13] for tool life testing in macro milling provides criteria and predictions exclusively for conventional cutting processes. The authors correlated cutting parameters with flank wear of a micro end mill using statistical treatments and found that wear depends highly on feed rate and tool diameter, and is directly associated with increased cutting force, part surface roughness, and tool cutting edge radius. They also proposed that flank wear could be used as a criterion for the tool life of micro end mills.

Therefore, as aforecited significant challenges in micromilling associated with chip formation, workpiece surface generation and tool wear monitoring still remain. Thus, this paper aims to quantify the wear/damages on the major and minor flank faces of the micro endmill cutter, identify the wear/damages mechanisms and their relation with shear-ploughing phenomena, and correlate the tool wear/damages and cutting strategy with the specific cutting force (size effect).

2 Methodology

Micromilling tests were carried out in a ROMI vertical CNC machining center model D600 adapted with a high-speed spindle NSK Nakanishi model NR3060S. The workpieces were prepared with dimensions of 10 × 26 × 58 mm from COS AR60 steel with an ultrafine-grained structure (UFG), 216 HV hardness and 0.7 μm average grain size. Carbide endmill tools from Sandvik Coromant (R216.32–00830-AI08G 1620) with 800 μm cutting diameter (D_c), two flutes, PVD TiAlN coating, and (1.9 ± 0.1) μm cutting edge radius (r_e) were used.

Feed per tooth (f_z) with two levels (3 and 10 μm/z) and cutting strategy with four levels (slot, symmetrical face, down and up milling) were considered as control factors (input variables) resulting in a combination of eight different cutting conditions. Thus, for each cutting strategy the following experimental criteria were adopted:

- Slot Milling ($a_e = D_c = 800$ μm): Simultaneous ploughing evaluation;
- Symmetrical Face Milling ($a_e = 0.625{\cdot}D_c = 500$ μm): Shearing dominant analysis;
- Down and Up Milling ($a_e = 0.375{\cdot}D_c = 300$ μm): Isolated ploughing investigation.

Each condition was replicated three times to perform statistical analysis by using Minitab® software version 21.1.0. Analysis of Variance (ANOVA) was applied by considering a significance level $\alpha = 5\%$. Cutting speed (v_c) and depth of cut (a_p) were kept constant at 60 m/min and 160 μm, respectively.

Figure 1 shows the experimental setup (left) and the cutting strategies (right). The tool followed a linear path on the xy plane and parallel to the y-axis. The active machining time of 1.63 min defined in pre-tests was adopted as tool-life criterion.

Wear and damage images were obtained using a Zeiss LEO 440 scanning electron microscope and tool flank wear magnifications were measured through AutoCAD® software version 24.3 by comparing with a new tool as a reference. The cutting force was measured employing a Kistler piezoelectric dynamometer model 9256C2. The xyz force components were post-processed using Matlab® version 2024b in order to calculate the specific cutting force [14].

3 Results and Discussion

Flank wear took place on both the major and minor flank faces. Figure 2 shows its magnification ranging between by about 30 and 120 μm.

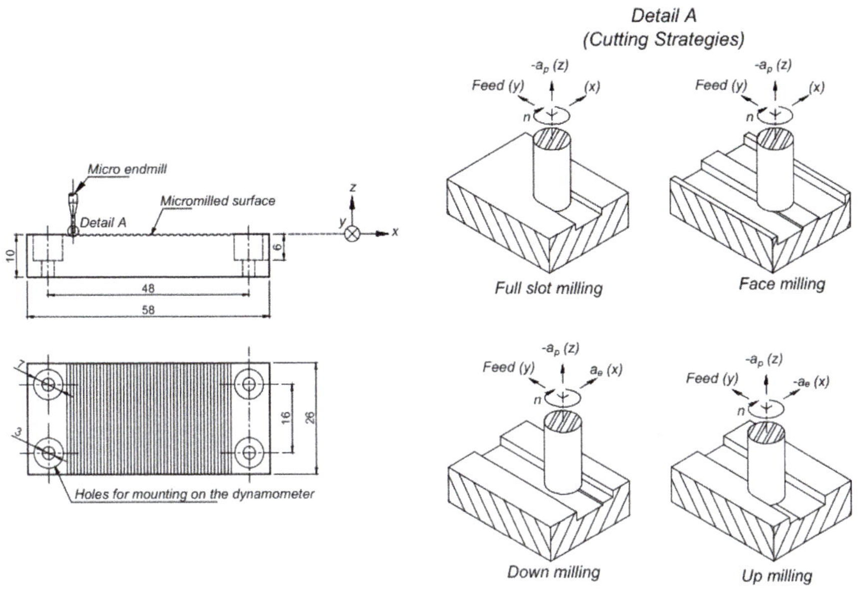

Fig. 1 Scheme of the experimental trials (left) and cutting strategies (right)

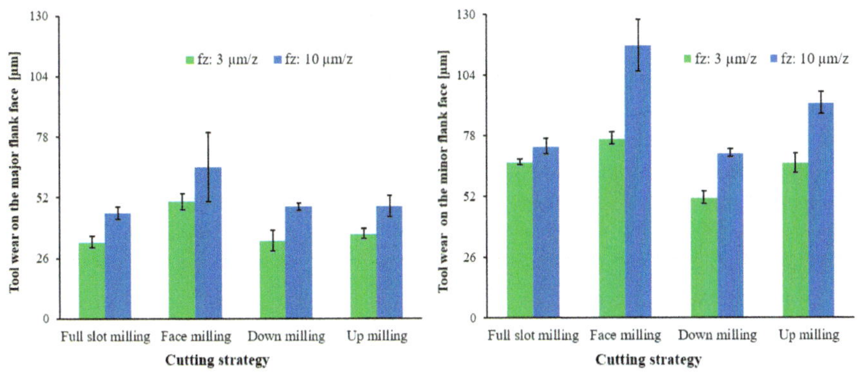

Fig. 2 Tool flank wear as a function of the tool feed and cutting strategy

It is noted that 10 μm/tooth tool feed increased the flank wear regardless of the cutting strategy and wear location most likely due to the greater cutting forces required to remove a larger chip volume. In this case the cutting force overcame the effect of the longer part-tool contact time for 3 μm/z feed per tooth. In particular for symmetrical face milling the tool-workpiece impacts chipped cutting edges being the main cause of tool damages.

A 233% increasing in feed per tooth augmented 35% the wear on both the major and minor tool flank faces. Down milling with a 3 μm/z feed per tooth exhibited the

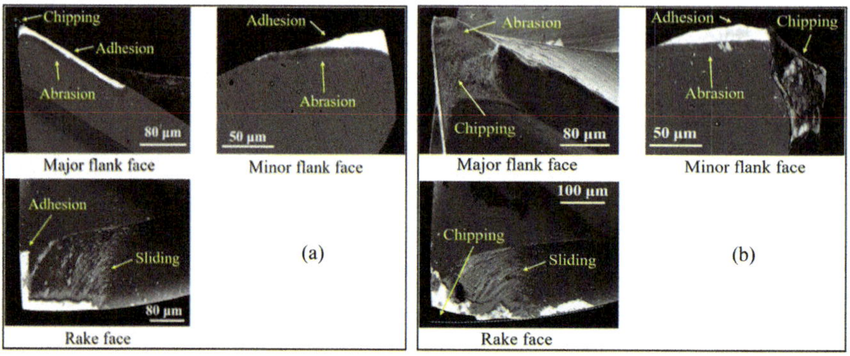

Fig. 3 Tool wear and damages for **(a)** 3 μm/z slot milling and **(b)** 10 μm/z face milling

best performance reaching lower wear in minor flank face. Minor flank face wear was on average 59% greater than that on the major flank face. This occurred likely due to the lesser tool clearance angle. The images showed in Fig. 3 support this qualitative discussion insofar as higher damages were not governed by ploughing but rather by edge-part impacts in each tool rotation (Fig. 3b).

For face milling with higher feed per tooth a significant material loss from the cutting edges due to chipping took place. Conversely, for lower feed per tooth and slot milling, tool wear caused by adhesion and abrasion mechanisms occurred (Fig. 3a).

Thus, under the tested conditions with lower feed per tooth a higher percentage of ploughing occurred along the tool path except for face milling. Instead of removing chips the ploughing effect otherwise compressed the workpiece material against the tool cutting edge radius causing as a consequence the tool flank wear. Additionally, the adhesion wear mechanism was intensified by variations in cutting thickness intrinsic of milling process. Therefore, when the ploughing effect is present during micro milling it likely promotes flank wear governed by adhesion and abrasion mechanisms.

Table 1 presents the ANOVA results which enables an assessment of how the effects of input variables and their first order interactions affect the response, specifically the tool flank wear on the major and minor flank faces.

It is identified that variations in feed per tooth and cutting strategy affect the behavior of flank wear on both the major and minor flank faces. *P*-value obtained for the control factors were lesser than the adopted significance level (5%). Thus, it can be stated with 95% confidence that only the main effects of each factor independently influenced the response. Therefore, flank wear variation directly depended on the feed per tooth or cutting strategy regardless of their first-order interactions.

Wear mechanisms mainly adhesion and abrasion given their incremental evolution can also be associated to size effect and part finishing or surface defects. Figure 4a shows the tool feed reduction increased the specific cutting force regardless of cutting strategy. This means the lower tool feed generated ploughing-dominant effect which

Table 1 ANOVA of the feed per tooth and cutting strategy on tool flank wear

Tool wear location	Control factors	Degrees of freedom	Sum of squares	Mean square	F-test	*P*-value
Major flank face	Feed per tooth	1	1080.01	1080.01	14.07	**0.002**
	Cutting strategy	3	1314.86	438.29	5.71	**0.007**
	Feed × strategy	3	10.93	3.64	0.05	0.986
Minor flank face	Feed per tooth	1	3154.02	3154.02	9.95	**0.006**
	Cutting strategy	3	4192.93	1397.59	4.41	**0.019**
	Feed × strategy	3	865.51	288.54	0.91	0.458

can be evidenced in Fig. 4b where the slot floor after micromilling revealed larger delaminated areas for 3 μm/z feed per tooth.

On the one hand for slot milling under 3 μm/z feed the ploughing effect was significantly more intense at the up milling side than that at the down milling side (Fig. 4b). This phenomenon augmented the tool flank wear as can be viewed in Fig. 2 especially at the minor flank face. On the other hand ploughing was minimized or even eliminated at the down milling side for higher feed per tooth since the cutting edge produces chips before reaches the minimum chip thickness. Burrs also identified in Fig. 4b were not the focus of this study.

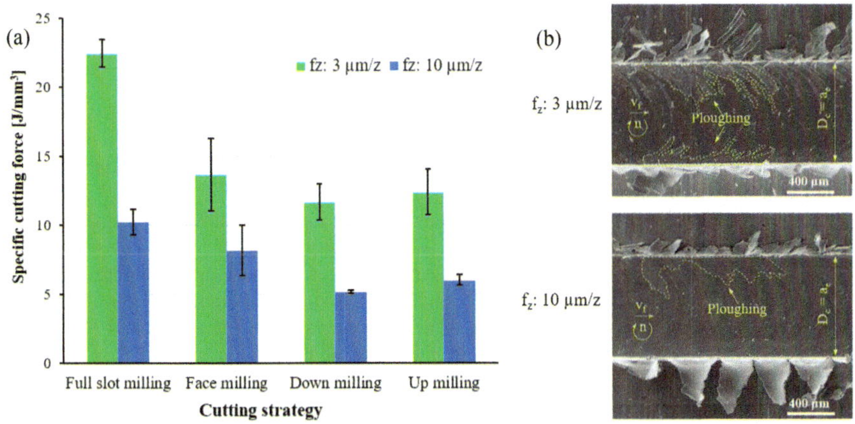

Fig. 4 **(a)** Specific cutting force as a function of tool feed and cutting strategy and **(b)** floor damages (delaminations) for slot milling strategy when varying tool feed

4 Conclusions

Based on the results of this study the following conclusions can be drawn:

- Considering variance analysis feed per tooth and cutting strategy were significant factors over the tool wear/damages on the major and minor flank faces;
- At higher feed per tooth the tool life is primarily caused by chipping mechanism. In contrast, at lower feed per tooth the dominant wear mechanisms were abrasion and adhesion;
- In slot milling and up milling strategies tool wear was governed by adhesion and abrasion mechanisms which in turn were powered by ploughing effect;
- Symmetrical face milling produced tool chipping damages due to the initial and final cutting thicknesses are larger than minimum chip thickness i.e., micromilling took place with impact-dominant and minimum or null ploughing;
- Increasing feed per tooth in 233% grows the tool flank wear on both major and minor flank faces in 35% on average;
- Flank wear on the minor flank face was about 59% greater than that on the major flank face;
- Down milling strategy at 3 μm/z feed yielded lower wear on the minor flank face;
- No difference was found for wear level on the major flank face when applying slot, down, and up milling strategies;
- Lower feed per tooth and slot milling strategy significantly increased the specific cutting force between 73% and 130%;
- Severe ploughed areas were produced at the up milling side and lower tool feed in slot milling strategy due to the size effect.

References

1. Kuram E (2024) Application of cutting fluids in micro-milling—a review. Int J Adv Manuf Technol 133:25–58
2. Balázs BZ (2021) A review on micro-milling: recent advances and future trends. Int J Adv Manuf Technol 112:655–684
3. Cristofaro S (2012) High speed micro-milling: novel coatings for tool wear reduction. Int J Mach Tools Manuf 63:16–20
4. Campos FO (2020) The influence of additive manufacturing on the micromilling machinability of Ti6Al4V: a comparison of SLM and commercial workpieces. J Manuf Process 60:299–307
5. Boswell B (2018) A review of micro-mechanical cutting. Int J Adv Manuf Technol 94:789–806
6. Masuzawa T (1997) Three-dimensional micromachining by machine tools. CIRP Ann 46(2):621–628
7. Mian AJ (2011) Identification of factors that dominate size effect in micro-machining. Int J Mach Tools Manuf 51(5):383–394
8. Cheng K, Huo D (2013) Micro-cutting: fundamentals and applications, 1st edn. Wiley, United Kington
9. Oliveira FB (2015) Size effect and minimum chip thickness in micromilling. Int J Mach Tools Manuf 89:39–54

10. Chae J (2006) Investigation of micro-cutting operations. Int J Mach Tools Manuf 46(3–4):313–332
11. Wang Z (2016) Built-up-edge effects on surface deterioration in micromilling processes. J Manuf Process 24:321–327
12. Colpani A (2019) Tool wear analysis in micromilling of titanium alloy. Precis Eng 57:83–94
13. International Organization Standardization (1989) ISO 8688: tool-life testing in milling
14. Rodrigues AR (2007) Influence of the tool edge geometry on specific cutting energy at high-speed cutting. J Braz Soc Mech Sci Eng 29(3):279–283

Application of Heat Map as Imaging Treatment to Enhance Surface Characteristics Visualization and Analysis for Milled Wire and Arc Additive Manufacturing Parts

Eduardo Neves Mota, Maksym Ziberov, Raphael Lima de Paiva, and Déborah de Oliveira

Abstract With the evolution of different additive manufacturing (AM) methods, it has become feasible to manufacture different types of parts, with complex geometries and different materials. For the manufacture of large metal parts, the process that stands out is the Wire and Arc Additive Manufacturing (WAAM), as it has high deposition rates, capacity to produce parts with different metal alloys and lower costs when compared to other AM deposition processes. WAAM parts, however, usually presents high roughness values and surface deviations, requiring post-processing operations. Among the post-processing methodologies, machining stands out in achieving specified shapes, dimensions and good finishing quality. Nevertheless, due to the non-homogeneous characteristics of the parts manufactured by WAAM, the cut can be hindered, and defects may occur during machining. In this sense, it is important to carefully analyze the surface and to differentiate machining from deposition irregularities on the surface. Therefore, this work aims to contribute to the surface evaluation of WAAM parts after the milling process. To aid in the visualization, the image was treated and converted into a heat map using open-source software Octave. This feature allows the transformation of the 2D image into a 3D topography, accentuating the surface characteristics. After evaluating the image and the generated topography, it is concluded that the combined use of the features favors a better surface topography evaluation of machined parts produced by WAAM.

Keywords Milling · WAAM · Surface quality · Surface evaluation · Heat map

E. N. Mota · M. Ziberov · D. de Oliveira (✉)
Department of Mechanical Engineering, University of Brasilia, Brasilia, Brazil
e-mail: oliveira.deborah@unb.br

R. L. de Paiva
School of Mechanical Engineering, University of Piauí, Teresina, PI, Brazil

M. C. Gomes et al. (eds.), *Innovations in Brazilian Machining*, Lecture Notes in Mechanical Engineering, https://doi.org/10.1007/978-3-031-92819-2_16

1 Introduction

Additive Manufacturing (AM) is a manufacturing method that allows the production of parts in layers and has become widespread due to its greater design freedom to produce different parts, which potentially leads to a reduction in weight, as well as in the number of production steps and costs [1]. In this method, the movements and trajectories of the deposition are determined through a 3D CAD model converted into layers to be deposited [2].

When focusing on metal parts there are more than 18 additive manufacturing methods, however, several of them become economically unfeasible due to the lack of technology available [3]. The ISO/ASTM 52900 standard restricts printing for metals and alloys to 4 methods such as Binder Jetting (BJT), Direct Energy Deposition (DED), Power Bed Fusion (PBF) and Sheet Lamination (SHL). However, for the manufacturing of metal parts, 3 types of techniques stand out [4], they are: PBF, BJT and DED.

The DED technique consists of launching a concentrated source of thermal energy in order to melt the filler material in the form of wire or powder into the pre-existing part. The ISO/ASTM 52900 standard defines the DED method as the process in which thermal energy is used to melt the materials as they are deposited. There are different DED processes, and they are associated with the manufacture of components that are not so complex and large in shape [5].

Among the different DED methods, the Wire and Arc Additive Manufacturing (WAAM) method is quite advantageous in economic aspects, presenting a cost of its raw material, the wire, a value of 10% of the powder cost, in addition to the possibility to use conventional welding machines as energy source, which makes the method cheaper than electron beam or laser methods. Another great advantage presented by WAAM method is its ability to generate fewer pores compared to powder-based methods, enhancing the quality of manufactured parts [6]. This is important to the fatigue life of the parts [7] which can be evaluated experimentally or numerically [8]. Different methods have been applied to improve WAAM parts, authors mention improvement when varying the deposition parameters and strategies [9], optimization due to the part geometry [10], and temperature control [11].

However, even with those improvements, post-processing is still necessary [12], and some authors report different difficulties when machining parts obtained by WAAM in comparison to parts made by casting or forging. Barreto [13] performed the milling of a ER70S-6 obtained by WAAM, with the aim of finding the best parameters to obtain a good surface finish. The milling was performed with a high-speed steel (HSS) cutter and the following parameters: cutting speed of 15.70 m/min, depth of cut of 0.25 mm and the feed per tooth varying as 0.005; 0.007; 0.010; 0.014 and 0.020 mm/tooth. For all the parameters, up milling and down milling were evaluated. For the condition in which a feed of 0.005 and 0.014 mm/tooth were applied, the surface presented a significant presence of pores, regardless of the cutting mode (up/down milling), and material detachment was also observed, as can be noticed in Fig. 1.

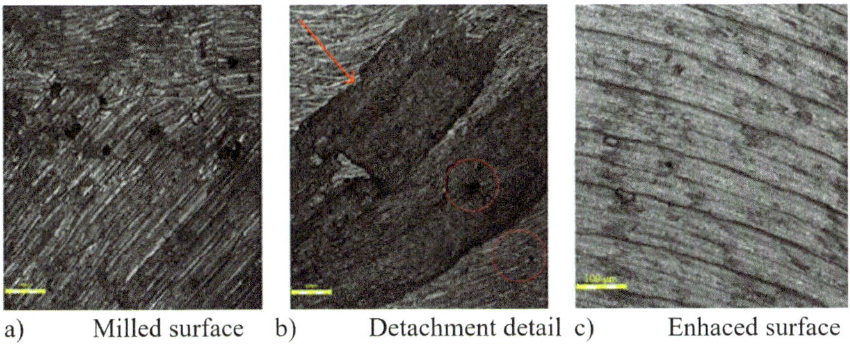

Fig. 1 Details of defects on surfaces of machined WAAM parts, **a**, **b** adapted from [13] and **c** from [14]

In Fig. 1a, it can also be noted that it is difficult to identify the machining marks and surface characteristics, since it presents marks in different directions, possible porosities and surface deformations. In another work, [14] performed machining a similar part, with the G9A69050 carbide mill with TiAlN coating. The authors evaluated different cooling-lubrication conditions in terms of number of nozzle, its positioning and cutting fluid flow rate in 20, 10 and 5 l/min leading to better surfaces when compared to the work of [13] as can be observed in Fig. 1c. It is noted that even with optimized cutting conditions, surface characteristics may still present variations/defects due to the deposition process, thereby making it difficult to properly analyze its topography through 2D images such as Fig. 1c. Thus, the objective of this work is to evaluate the use of the image treatment through heat map, to improve the visualization of the surface characteristics of parts produced by WAAM, improving surface analysis after the milling process.

2 Methodology

The manufacturing of the machined parts was based on the work of [15]. The authors studied the geometry of walls manufactured by WAAM, using an ER70s-6 wire. The material was then machined on the XH7132 CNC, with 3.7 kW. The cutting tool was a carbide mill with 5 mm diameter, 4 cutting edges and (Ti, Al)N coating. Based on the work of [14], the BIO100e cutting fluid, a biodegradable synthetic oil, from the manufacturer Biolub Química Ltda, was applied with a flow rate of 10 l/min. The cutting parameters are presented in Table 1.

Since the objective is to investigate the use of computational resources for surface evaluation, after machining it is necessary to obtain images of the surfaces through microscopy. For this, an Olympus LEXT OLS4100 microscope was used. Images of the machined surface, with different magnifications and in different regions, were obtained. Once the image was saved, in high quality, it was processed. For this

Table 1 Cutting parameters

Parameter	Value
Depth of cut	0.25 mm
Feed per tooth	0.02 mm
Radial depth of cut	5.00 mm
Cutting speed	10 m/min
Machining length	30.00 mm

purpose, the open-source software Octave was used, and the programming included the following steps:

i. Inserting the image into the software;
ii. Transforming the image into a pixel matrix;
iii. Cutting the matrix to the region to be analyzed;
iv. Assigning the gray scale value to each pixel;
v. Generating a numerical matrix;
vi. Normalizing the values;
vii. Applying the heat map function to the numerical matrix;
viii. Generating topography images.

3 Results and Discussions

Figure 2 contains a microscopic image of the machined surface. It shows that there are several characteristics that need to be observed after milling. Figure 2a represents a region of the machined surface that may have porosities due to the WAAM process. Figure 2b shows a broader view of the slot. In addition to the pores, surface stains and feed marks can be seen. When looking in detail at Fig. 2c, it is possible to observe that there are dark spots and that due to surface variations, the feed marks become difficult to differentiate.

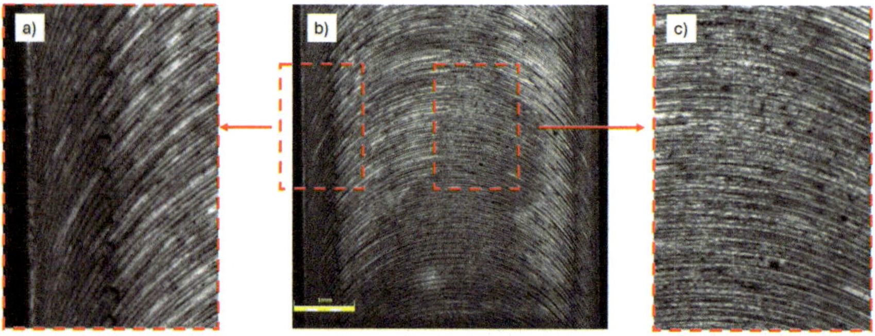

Fig. 2 Surface of MADA workpiece after milling, image obtained with 2.5x magnification

To improve visualization, Fig. 3 contains images obtained via microscopy with higher magnifications, Fig. 3a 10.0x and b 5.0x. Note that, in Fig. 3b, the feed marks can be clearly seen, but without sufficient sharpness to observe possible defects or small plowing. Usually, higher magnifications are used to observe small details, such as those mentioned, however, one can note in Fig. 2a, that the variations in the image do not allow the correct focus of the microscope, generating an unclear image, for the same acquisition configuration.

It is therefore necessary to improve the visualization of the surfaces of parts obtained by WAAM, to allow an adequate analysis of the process. Thus, Fig. 4 shows the application of image processing by applying the heat map as the methodology proposed in this work.

It is possible to note that the use of the heat map favors the visualization of surface details, highlighting, for example, irregularities in the feed marks, which in Fig. 4a could be confused with other superficial defects such as pores, adhered material and

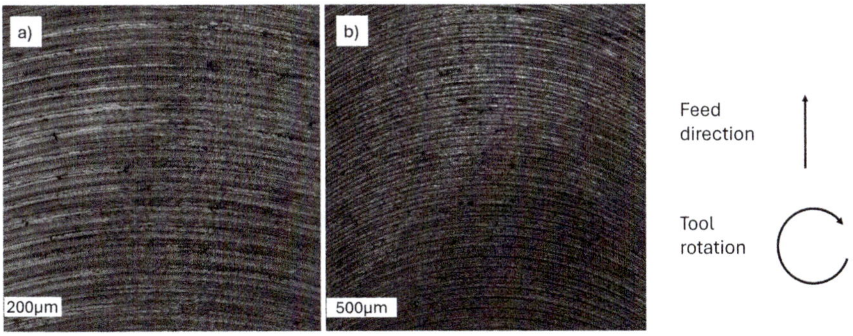

Fig. 3 Surface of WAAM part after milling, **a** 10.0x magnification and **b** 5.0x magnification

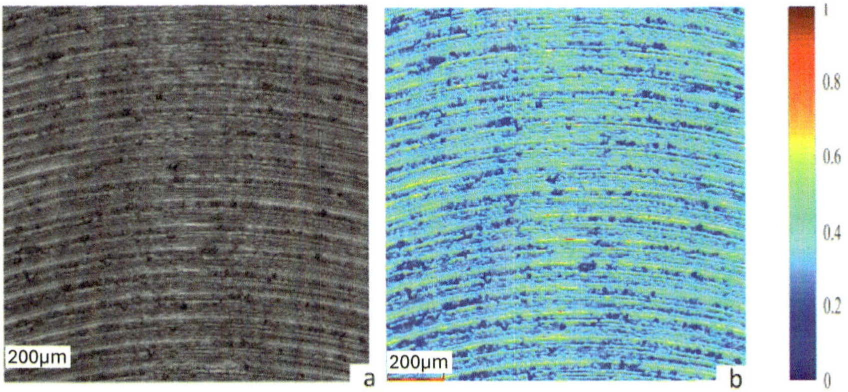

Fig. 4 Surface of the WAAM part after milling, **a** as obtained by microscopy and **b** after image processing

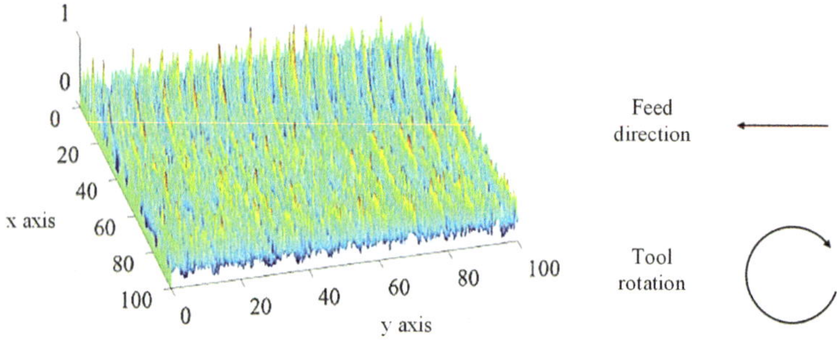

Fig. 5 Topography of the WAAM part after milling

intense plastic deformation. These topography variations/defects may be attributed to the microstructure that WAAM parts present, which is a result of large thermal gradients during the deposition process [11]. Lopes et al. [16] observed similar microscopy images after milling the high strength low alloy (HSLA) steel produced by WAAM with uncoated carbide end mill. Besides feed marks, the authors detected superficial defects on the machined surface, which were specified as adhered material, pronounced grooves, and material side flow. According to the authors, adhered material is associated with build-up edge (BUE) formation, while the formation of pronounced grooves occurs randomly due to the increase friction between the tool and the workpiece material. Material side flow, on the other hand, is related to plastic deformation with no material removal.

Finally, computational resources also allow the obtaining of three-dimensional topography as shown in Fig. 5. This type of representation can favor the understanding/evaluation of the machined surface topography, especially considering feed marks and superficial defects, and it can be applied for a better comparison between different cutting conditions with a higher level of detail in parts manufactured by grinding process [17].

4 Conclusions

After machining parts obtained by WAAM, and analyzing the microscopy images and heat maps, the following conclusions could be drawn:

i. The surface analysis of parts obtained by WAAM can be difficult due to their characteristics such as anisotropy, large variations in the microstructure and presence of defects.
ii. The use of computational resources associated with microscopy images through heat map plots contributes to analyze the surface topography of the part after milling by highlighting feed marks and/or surface variations/defects resulted from WAAM process.

Acknowledgements The authors are grateful to the University of Brasilia and CNPq, grant numbers 402730/2023-2, 443148/2023-6 and 200018/2025-6 for the financial support, and LAETA/INEGI—Associate Laboratory of Energy, Transports and Aerospace/ Institute of Science and Innovation in Mechanical Engineering and Industrial Engineering.

References

1. Cotteleer M, Joyce J (2014) 3D opportunity: additive manufacturing paths to performance, innovation, and growth. Deloitte Rev 14(1):3–19
2. Gibson I, Rosen DW, Stucker B (2010) Additive manufacturing technologies. Rapid prototyping to direct digital manufacturing. Springer, New York, NY
3. Thomas-Seale LEJ, Kirkman-Brown JC, Attallah MM, Espino DM, Shepherd DET (2018) The barriers to the progression of additive manufacture: perspectives from UK industry. Int J Prod Econ 198:104–118
4. Toyserkani E, Dyuti S, Ibhadode OO, Liravi F, Russo P, Taherkhani K (2021) Metal additive manufacturing, 1st edn. Wiley, NJ
5. Debroy T, Wei H, Zuback J, Mukherjee T, Elmer J, Milewski J, Beese A, Wilson-Heid A, De A, Zhang W (2018) Additive manufacturing of metallic components—process, structure and properties. Prog Mater Sci 92:112–224
6. Lee SH (2020) Optimization of cold metal transfer-based wire arc additive manufacturing processes using Gaussian process regression. Metals 10(4):461
7. Araujo LC, Ferreira JLA, Ziberov M, Araujo JA (2023) Assessing fatigue in materials with small defects: a new multiaxial model based on principal stress amplitudes. Procedia Struct Integr 57:144–151. https://doi.org/10.1016/j.prostr.2024.03.017
8. Silva CP, Silva GP, Santos MC, Ziberov M, Malcher L (2024) Hybrid search methodology for mechanical characterization of material produced via WAAM assuming Gurson porous material. J Braz Soc Mech Sci Eng 46(266). https://doi.org/10.1007/s40430-024-04828-8
9. Vieira RDAM, Liskevych O, Ziberov M (2024) Influence of deposition strategy in CMT-based wire and arc additive manufacturing on part geometry: a literature review. Soldagem Inspeção 29(e2910). https://doi.org/10.1590/0104-9224/SI29.10
10. Vieira RDAM, Liskevych O, De Oliveira D, Ziberov M (2024) Influence of parameter variation and interlayer temperature control in wall angle, curvature and measurement methodology of ER70S-6 parts obtained by WAAM. Manuf Lett 42:40–45. https://doi.org/10.1016/j.mfglet.2024.10.006
11. De Araújo Soares MA, Novelino ALB, Ziberov M (2024) Geometry study on 410NiMo alloy parts printed by WAAM-CMT. In: de Oliveira D, Ziberov M, Rocha Machado A (eds) ABCM series on mechanical sciences and engineering. COBEF 2023. LNCS, pp 114–125. Springer, Cham. https://doi.org/10.1007/978-3-031-43555-3_11
12. De Oliveira D, Gomes MC, Dos Santos AG, Ribeiro KSB, Vasques IJ, Coelho RT, Da Silva M, Hung WNP (2023) Abrasive and non-conventional post-processing techniques to improve surface finish of additively manufactured metals: a review. Prog Addit Manuf 8:223–240. https://doi.org/10.1007/s40964-022-00325-3
13. Barreto TSM (2022) Estudo das características de usinagem de peças metálicas obtidas por manufatura aditiva a arco, Monograph. Department of Mechanical Engineering, University of Brasilia, Brasilia, pp 1–80
14. Ribeiro GM, Goncales MV, de Oliveira D (2023) Influence of flow and positioning of the nozzle of the cutting fluid in the milling of ER70s-6 parts obtained by WAAM. In: 12th Brazilian manufacturing engineering congress. Brasilia, pp 1–9
15. Novelino ALB, Carvalho GC, Ziberov M (2022) Influence of WAAM-CMT deposition parameters on wall geometry. Adv Ind Manuf Eng 5:100105. https://doi.org/10.1016/j.aime.2022.100105

16. Lopes JG, Machado CM, Duarte VR, Rodrigues TA, Santos TG, Oliveira JP (2020) Effect of milling parameters on HSLA steel parts produced by wire and arc additive manufacturing (WAAM). J Manuf Process 59:739–749
17. De Paiva RL, De Oliveira D, Ruzzi RS, Jackson MJ, Gelamo RV, Abrão AM, Da Silva RB (2023) Surface roughness and morphology evaluation of bearing steel after grinding with multilayer graphene platelets dispersed in different base fluids. Wear 523:204852. https://doi.org/10.1016/j.wear.2023.204852

Machining of Hybrid Rocket Engine Nozzle Manufactured by Wire and Arc Additive Manufacturing

Rafael Silva Barros, Déborah de Oliveira, Olexiy Shynkarenko, Antônio Piratelli-Filho, and Maksym Ziberov

Abstract Due to its ability to reduce manufacturing costs and time, and the possibility of manufacturing parts with complex geometry, additive manufacturing (AM) in metals is attractive for the industry. The wire and arc additive manufacturing (WAAM) can produce large parts with high deposition rate, surpassing other AM processes. However, for the WAAM process to become truly suitable for engineering applications, it is necessary for the parts to undergo post-processing, since they have an irregular surface. In this sense, this work aims to validate the machining strategy of a hybrid rocket engine nozzle produced by the WAAM. The part was initially manufactured in polymer by material extrusion, with the geometry modified according to the proposed machining strategy. The turning was then performed on a conventional lathe to verify the proposed geometry. The same methodology was applied to a metal part manufactured via WAAM. After the machining and analysis, it was possible to observe that for the polymer part, the validation was satisfactory, once it was possible to obtain the design geometry. During the deposition of the metal part, the heat input caused angular deformation of the substrate, which would be fixed to the lathe, hindering the positioning, and, therefore, there was a variation in the wall thickness of the part at the end of the machining.

Keywords WAAM-CMT · Machining strategy · Turning · Surface finishing · Aerospace application

R. S. Barros · D. de Oliveira · A. Piratelli-Filho · M. Ziberov (✉)
Department of Mechanical Engineering, University of Brasilia, Brasilia, Brazil
e-mail: mziberov@unb.br

O. Shynkarenko
Department of Aerospace Engineering, University of Brasilia, Brasilia, Brazil

M. C. Gomes et al. (eds.), *Innovations in Brazilian Machining*, Lecture Notes in Mechanical Engineering, https://doi.org/10.1007/978-3-031-92819-2_17

1 Introduction

The aerospace industry has been growing significantly in recent years, with the space sector expected to grow by around US$1 trillion by 2040 [1]. Due to this growth, the need for new technologies that reduce fuel consumption [2] and lighter parts that aim to optimize weight are also growing [3]. Some aerospace materials, however, have low machinability, increasing manufacturing costs due to tool wear and material waste [4]. In this context, it is important to optimize production by investing in alternative processes to traditional manufacturing, such as additive manufacturing (AM).

Wire and arc additive manufacturing (WAAM) stood out due to its ability to produce parts with complex geometries, high deposition rates, lower equipment costs, and use of different types of metals [5]. Among the technologies used in with WAAM, Cold Metal Transfer (CMT) stands out once it has greater control of the deposition, lower thermal input, leading to more precise geometry [6] which can also be combined with the deposition strategy to further improvements [7].

Within the characteristics, the part geometry stands out. Soares et al. [8] analyzed the geometry of ER410NiMo alloy parts deposited by WAAM-CMT. The deposition of a wall with temperature control presented an effectiveness ratio of 86%, being 14% greater than a wall deposited without control. This is even more important when depositing inclined parts, since geometry will have great influence on the usefulness of the deposited part [9]. Due to this, it is necessary to understand post-processing [10] to optimize the machining parameters for parts produced by AM [11].

In this sense, the AM project must consider machining aspects such as the fixation of the part in the machine tool, as well as the trajectory of the tool [12]. Furthermore, depending on the application of the part, it is necessary to adequately combine the additive and subtractive manufacturing, and the safety of the operation. Considering a specific application, this work will evaluate the machining strategy of a hybrid rocket engine nozzle, Fig. 1, manufactured via WAAM-CMT. Therefore, this work aims to validate the nozzle machining strategy. The validation took place in two stages, one of them in a part produced by polymer additive manufacturing and the other in a metal part obtained by the WAAM-CMT process.

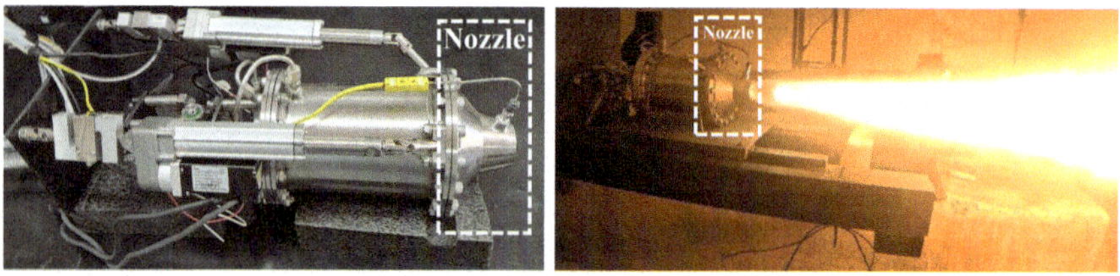

Fig. 1 Hybrid rocket engine test model

2 Materials and Methods

To evaluate the machining strategy, as a preliminary trial, a polymer part was produced using the Sermoon V1 Pro Creality 3D Printer. The material used in the deposition was PLA filament, with a diameter of 1.75 mm. For the studied part, the metal deposition was performed using a combination of a welding power source and a torch fixed to a Cartesian robot. The Cartesian robot used was the Schneider Electric MAXR23-S42-H42-C42 and the power source used was the Fronius TransPuls Synergic 5000 CMT, with a working volume of 800 × 800 × 500 mm and an accuracy of 0.10 mm. The material used was ER70s-6 wire with 1.20 mm in diameter, whose chemical composition is 0.07%C, 0.85%Si and 1.50%Mn (by weight). The shielding gas used was the Ar+18%CO_2 mixture with a flow rate of 15–20 l/min.

As part of the machining strategy, the correct preparation of the substrate is very important. A SAE 1020 carbon steel substrate with dimensions of 200 × 200 × 6.40 mm was used. The center of the part was marked, and the deposition torch was positioned after fixing the substrate to the Cartesian system and the trajectory test was performed. This is important to avoid decentralization when the part was fixed to the lathe.

The machine tool used was the conventional IMOR RN-400 mechanical lathe, usually present in Brazilian universities [13]. The cutting tool used for external machining was the TNMG 160404E-FM (tool radius of 0.40 mm) from Dormer Pramet, for internal machining the tool was the ISO 8 10 × 10 DP 30 from Rocast, with a 90° tip angle. The cut was performed dry considering the Sustainable Development Goals (SDG) 9 which deals with industry, innovation and infrastructure and 12 which addresses responsible consumption and production [14]. The machining parameters in the longitudinal direction were kept as follows: rotation (n) = 200 rpm, feed (f) = 0.072 mm/revolution, depth of cut (ap) = 0.30 mm and feed speed (vf) = 14.40 mm/min. After machining, all the chips generated were collected and weighed using a digital scale with a capacity of 10 kg and accuracy of 1 g, to obtain the ratio between deposited and machined material. After machining, roughness measurement was performed using a Mitutoyo Model SJ-201 portable surface roughness tester with a cut-off of 0.8 mm, the part was measured in four regions, for each one 5 measurements were made and the results were calculated as the values with standard deviation.

3 Results

The polymer printing lasted approximately 28 h. Figure 2a shows the 2D drawing of the nozzle with the design geometry (height of 65 mm and diameter of 153 mm). The upper and lower conical regions are highlighted, which have angles of 73° and 40.50° respectively, with a constant wall thickness of 5 mm.

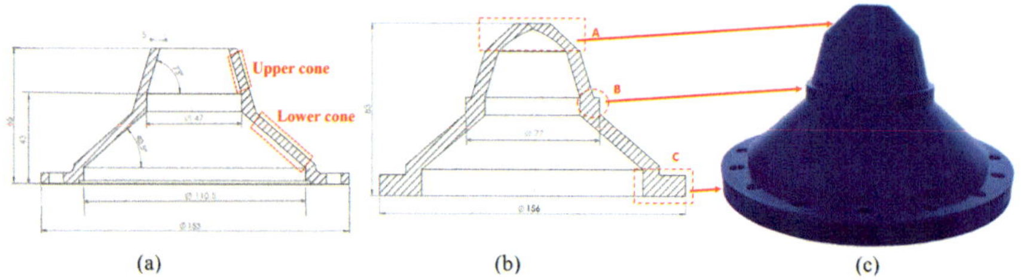

Fig. 2 **a** 2D drawing with design geometry, **b** 2D drawing with geometry after manufacturing and **c** real part manufactured by material extrusion

Figure 2b shows the part with the appropriate changes that aid in machining, Fig. 2c shows the actual part manufactured in polymer according to the 2D drawing. The points (A, B and C) highlighted in Fig. 2b are temporary support points with the following purposes: support point A is where the center hole will be drilled, allowing the part to be supported by a tailstock. Support point B is where the part will be fixed during internal machining. Support point C represents the additional amount of material that will be deposited at the base of the part, allowing safe extraction of the substrate and ensuring sufficient excess metal for finishing in that region.

The machining followed the steps: (a) fixing to the lathe through the substrate, using a plate with 4 independent jaws; (b) machining of support points B and C, i.e., leveling and alignment; (c) after removing the substrate, the part was fixed to the lathe through support point B. In this step, support point C was machined with the design dimensions; (d) using a cutting tool for internal turning, the lower conical region was machined. It was necessary to tilt the upper slide at 40°; (e) removal of support point A, through the radial bleeding operation; (f) using an external tool, the part was faced; (g) using an internal cutting tool, the upper cone was machined to the final design dimension. The upper slide had to be tilted at 17° and (h) support point B was removed, and the upper and lower cones were externally machined, requiring the upper slide to be tilted at 73° for machining the upper cone and tilted at 40° for machining the lower cone.

For the deposition of the metal part, two different deposition parameters were used, as shown in Table 1. For the upper cone and lower cone regions of support point B, the parameters were milder due to the angles [9] and for support point C, a greater heat input is desired, so that there is fusion between the layers once there is more material deposited to the sides [15]. For this purpose, a current 37.30% higher was used in relation to the current of the upper cone and lower cone.

Figure 3a shows the part at the end of the WAAM-CMT deposition. Support point B, Fig. 3b, consists of the deposition of single beads/layers deposited successively, with an average diameter of 77.20 mm and a height of 23 mm. Three thickness measurements were made in this region, with an average of 4.60 mm. The deposition of the lower conical region, Fig. 3c, was performed in a similar manner, also with single beads/layers, totaling 66 layers, which were deposited in three blocks, to adjust the contact tip-to work distance, with diameter varying from 112.90 to 77.20 mm and

Table 1 Nozzle deposition parameters

Deposition region	Current (A)	Wire feed speed (m/min)	Torch travel speed (mm/s)	Contact tip-to work distance (mm)	Step-over distance (mm)*
Lower and upper cone region [9]	69	1.40	6	12	1.20
Support point region C [15]	110	3.40	8		2.50

*Step-over distance is vertical up movement at the end of each deposited layer

a height of 37 mm. The inclination of the conical region presented an approximate angle of 64°, different from the design angle of 40°. Support point C, Fig. 3d, consists of a filled "ring", with an outer diameter of 156 mm, an inner diameter of 100 mm, and a height of 30 mm. In this case, the deposition occurred satisfactorily, as it has sufficient excess metal to achieve the geometry as shown in Fig. 3a.

The machining strategy previously validated on the polymer part was applied to the metal part until the layer structure could no longer be observed. Thus, considering the points addressed in the validation, the machining of the metal part followed the steps: (a) fixing the part in the lathe through the substrate; (b) machining of support points B and C, so that these points are parallel to the center line of the part, being reference for the following steps and (c) partially machined part, with substrate.

The substrate could not be removed on the lathe as planned; thus, the removal was done with plasma cutting, using a HandyPlasma 45i ESAB machine, and a Manrod MR105 horizontal band saw. The other steps were performed after the substrate was removed. Figure 4a presents the machining of the support point C using a facing operation. Figure 4b represents the machining of the lower conical region. In this step, it is necessary to tilt the upper slide at 64°; the feed was done manually and Fig. 4c the final part at the end of the machining process.

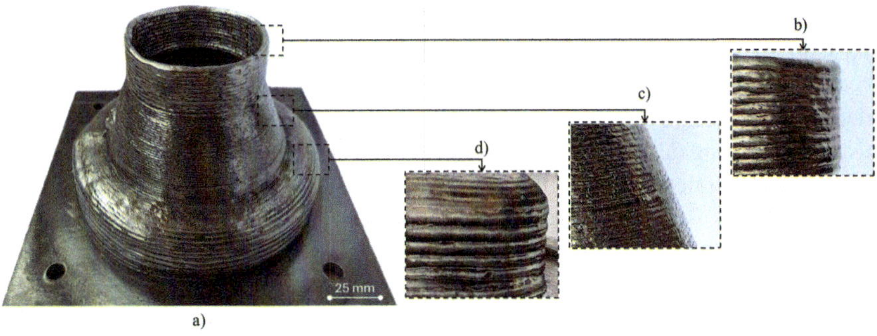

Fig. 3 **a** Part at the end of the WAAM-CMT process, **b** detail of support point B, **c** detail of the lower conical region and **d** detail of support point C

(a) (b) (c)

Fig. 4 Application of the machining strategy to the metal part after removing the substrate

The deposition of the support point C was made in 8 layers and in each layer 14 circular layers were deposited, with 2 mm between each one. During this deposition it was necessary to maintain the current at 110 A, to guarantee homogeneity between the layers. However, the heat input generated caused an angular deformation of 9.50° in the substrate, to mitigate this problem one can choose a thicker substrate and/or better fixation. The total mass of the part before machining was 2576 g, and 906 g of material was removed through machining, which corresponds to 35.20%. Excess material was deposited in the region of support point C, where the largest amount of material was machined; these parameters can be adjusted in future work.

A direct consequence of the decentralization is that the machining occurred unevenly, and this can be clearly observed in Fig. 5. At support point B, the average wall thickness before machining was 4.59 mm (WT), Fig. 5a. Since the machining occurred eccentrically, support point B presented an uneven structure, with the wall thickness varying from 2.65 mm at the thinnest point to 4.10 mm at the thickest point, Fig. 5b. It is worth mentioning that this was observed at support point B because it is the easiest point to visualize, but, given the nature of the turning process, this characteristic should also be found in the other regions. It can be seen that the difference between the wall measurement before machining and the largest measurement after machining is approximately 0.49 mm. It can be inferred that the effective width in the region of support point B would be approximately 3.61 mm, resulting in a WEf/WT ratio of 78%. This value can be considered satisfactory, since it would be necessary to machine approximately 0.98 mm of the wall to obtain WEf. However, this is an ideal value, which does not consider factors such as misalignment of the part and distortions during deposition, for example. When considering such factors, it would be necessary to deposit more material, which would result in a larger WT and a smaller WEf/WT ratio, but which would meet the quantity of material sufficient to machine the part with the design dimensions.

After machining the part, it was possible to observe the presence of some discontinuities, such as the presence of pores (micropores, i.e., discontinuities with smaller size/area) and points where there was a lack of fusion between layers, shown highlighted (red arrows) in Fig. 5c. These discontinuities appeared more present in the

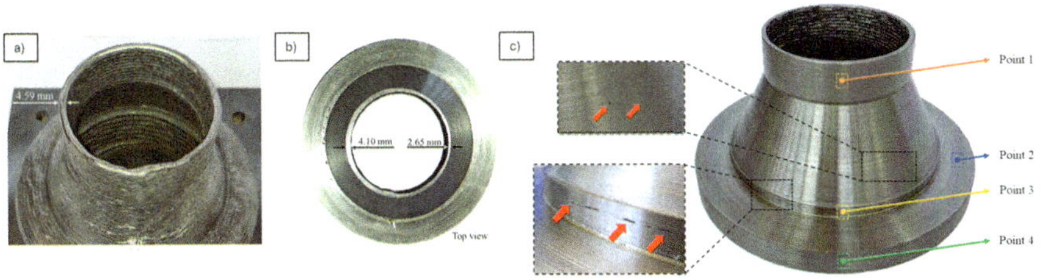

Fig. 5 **a** Wall thickness before machining, **b** wall thickness after machining and **c** quality on different surface regions

region of points 1, 2, and 3 in Fig. 5c where deposition parameters with lower heat input were used. According to Zeng et al. [16], due to repeated thermal cycles, parts produced by WAAM are subject to some inherent defects of the process, such as porosity, cracks, incomplete fusion, delamination, oxidation, and residual stress. The discontinuities can be described by Gurson's mathematical model [17]. Those methods are important once to comprehend the material behavior and to estimate the reduction on the fatigue life of parts with small defects [18].

To roughness values can be observed in Fig. 6, measured in the points indicated in Fig. 5c. At points 1, 3 and 4, machining occurred in the direction parallel to the axis of the workpiece, at point 2, the facing operation was performed, machining in the direction perpendicular to the axis of the workpiece. The highest roughness values were found at points 2 and 3, where the tool feed was performed manually. Similar Ra results were obtained by Rosa et al. [19], where SAE 1020 and SAE 1045 steel were machined by turning with the parameters of vc = 100 m/min, ap = 1 mm and f = 0.16 mm/rev. The values of Ra = 3.73 μm were obtained for SAE 1020 steel and Ra = 2.32 μm for SAE 1045 steel.

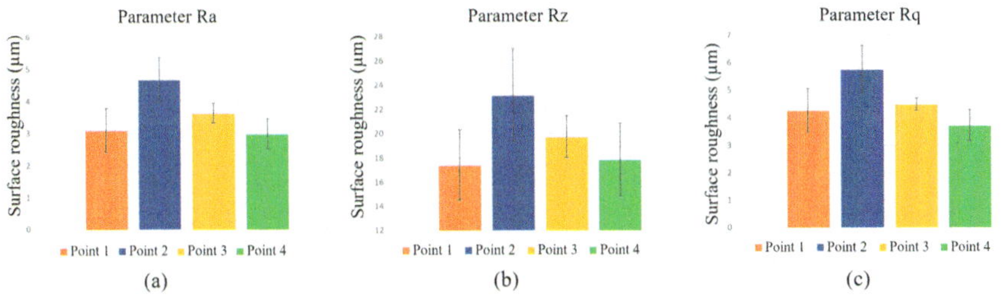

Fig. 6 Surface roughness, **a** Ra values, **b** Rz values and **c** Rq values

4 Conclusions

After the study of the turning of the WAAM part, the following conclusions could be drawn: It is important to consider the over metal for machining and the addition on the CAD model of the support points, enabling the machining; Applying the strategy to a polymer part can contribute to the machining strategy quality, to verify of the fixation points and the ability to achieve the final geometry; The angular deformation of the substrate during the deposition of the metal part can lead to wall thickness variation after machining; Surface roughness varied from 2.98 to 4.67 μm (Ra), similar to the literature for low alloy steels, demonstrating that the machining parameters used for those steels can be applied to parts produced with ER70S-6 wire via WAAM-CMT.

Acknowledgements The authors are grateful to the University of Brasilia and CNPq, grant numbers 405499/2022-1, 402730/2023-2, 443148/2023-6 and 200018/2025-6 for the financial support, and LAETA/INEGI—Associate Laboratory of Energy, Transports and Aerospace/ Institute of Science and Innovation in Mechanical Engineering and Industrial Engineering.

References

1. Exame, https://exame.com/invest/mercados/industria-aeroespacial-vai-movimentar-1-trilhao-nos-proximos-anos-aqui-esta-o-porque/. Last accessed 06 Jun 2024
2. Silva HL, Guimarães TAM, Ziberov M (2022) Hybrid-electric airplane design optimization considering climb and cruise phases. J Braz Soc Mech Sci Eng 44(41). https://doi.org/10.1007/s40430-021-03311-y
3. Wickramasinghe KC, Sasahara H, Rahim EA, Perera GIP (2021) Recent advances on high performance machining of aerospace materials and composites using vegetable oil-based metal working fluids. J Clean Prod 310:127459
4. Mishra SK, Ghosh S, Aravindan S (2020) Machining performance evaluation of Ti6Al4V alloy with laser textured tools under MQL and nano-MQL environments. J Manuf Process 53:174–189
5. Wu B, Pan Z, Ding D, Cuiuri D, Li H, Xu J, Norrish J (2018) A review of the wire arc additive manufacturing of metals: properties, defects and quality improvement. J Manuf Process 35:127–139
6. Novelino ALB, Carvalho GC, Ziberov M (2022) Influence of WAAM-CMT deposition parameters on wall geometry. Adv Ind Manuf Eng 5:100105. https://doi.org/10.1016/j.aime.2022.100105
7. Vieira RDAM, Liskevych O, Ziberov M (2024) Influence of deposition strategy in CMT-based wire and arc additive manufacturing on part geometry: a literature review. Soldagem Inspeção 29(e2910). https://doi.org/10.1590/0104-9224/SI29.10
8. De Araújo Soares MA, Novelino ALB, Ziberov M (2024) Geometry study on 410NiMo alloy parts printed by WAAM-CMT. In: de Oliveira D, Ziberov M, Rocha Machado A (eds) ABCM series on mechanical sciences and engineering. COBEF 2023. LNCS. Springer, Cham, pp 114–125. https://doi.org/10.1007/978-3-031-43555-3_11
9. Vieira RDAM, Liskevych O, De Oliveira D, Ziberov M (2024) Influence of parameter variation and interlayer temperature control in wall angle, curvature and measurement methodology of ER70S-6 parts obtained by WAAM. Manuf Lett 42:40–45. https://doi.org/10.1016/j.mfglet.2024.10.006

10. Gomes MC, Dos Santos AG, De Oliveira D, Figueiredo GV, Ribeiro KSB, De Los Rios GAB, Da Silva MB, Coelho RT, Hung WNP (2022) Micro-machining of additively manufactured metals: a review. Int J Adv Manuf Technol 118:2059–2078. https://doi.org/10.1007/s00170-021-08112-0
11. De Oliveira D, Gomes MC, Dos Santos AG, Ribeiro KSB, Vasques IJ, Coelho RT, Da Silva M, Hung WNP (2023) Abrasive and non-conventional post-processing techniques to improve surface finish of additively manufactured metals: a review. Prog Addit Manuf 8:223–240. https://doi.org/10.1007/s40964-022-00325-3
12. Pittman K, https://www.engineering.com/optimizing-machining-and-workholding-for-metal-additive-manufacturing/. Last accessed 12 Jun 2024
13. Santiago LCB, Nunes BS, Amaral IR, Conceição MS, Dos Santos AG, De Oliveira D (2022) Maintenance plan proposal for a turning machine from machining laboratory of university of Brasilia. Braz J Prod Eng 8(4):132–152. https://doi.org/10.47456/bjpe.v8i4.38701
14. Sustainable Development Goals (SDGs), https://www.undp.org/sustainable-development-goals. Last accessed 01 Aug 2024
15. Novelino ALB (2023) Fadiga Multiaxial em Peças Fabricadas via Manufatura Aditiva por Deposição a Arco via CMT. Dissertation (Master in Mechanical Sciences), University of Brasilia, Brasilia, pp 1–96
16. Zeng Y, Wang X, Qin X, Hua L, Xu M (2010) Laser ultrasonic inspection of a wire + arc additive manufactured (WAAM) sample with artificial defects. Ultrasonics 110:106273
17. Silva CP, Silva GP, Santos MC, Ziberov M, Malcher L (2024) Hybrid search methodology for mechanical characterization of material produced via WAAM assuming Gurson porous material. J Braz Soc Mech Sci Eng 46(266). https://doi.org/10.1007/s40430-024-04828-8
18. Araujo LC, Ferreira JLA, Ziberov M, Araujo JA (2023) Assessing fatigue in materials with small defects: a new multiaxial model based on principal stress amplitudes. Procedia Struct Integr 57:144–151. https://doi.org/10.1016/j.prostr.2024.03.017
19. Rosa AFP, Hammes JM, Royer R (2020) Surface roughness optimization of SAE 1020 and SAE 1045 turning process using design of experiments. Braz J Dev 6(10):79069–79087

Effect of the Wire Electrical Discharge Machining and Electropolishing on Surface Quality of the Nd-Fe-B Magnet

Erick Cardoso Costa, Gustavo Sorgetz, Natalie Souza Heinz, Beatriz Cardoso Santos, and Fabio Antonio Xavier

Abstract Nd-Fe-B-based permanent magnets are essential for a wide range of industrial, military, and consumer applications. Machining processes are crucial to their production, as they ensure precise dimensions, accurate shapes, and high-quality surface integrity, while also maintaining tight tolerances. However, despite their importance, the scientific literature lacks comprehensive information on the machining of Nd-Fe-B magnets, especially regarding the surface formation mechanisms involved in processes including wire electrical discharge machining (WEDM). In this study, Nd-Fe-B samples were first machined using WEDM under two finishing conditions (Finishing A and B) and then subjected to electropolishing using a nitric acid solution in denatured alcohol. The process involved varying the current density and processing time. The surface quality was analyzed by measuring surface morphology and surface roughness (R_a and R_q), and the material removal rate (MRR) was calculated. The results obtained showed that electropolishing significantly improved surface smoothness, especially at higher current densities and longer processing times. In this case, increasing the current density to 1.0 A/cm^2 and extending the processing time to 10 min led to a notable reduction in surface roughness, with the best results observed under Finishing A. Additionally, the material removal rate increased with both current density and time, the surface morphology from WED-machining in Finishing B enhances material removal during electropolishing. The MRR was significantly influenced by current density and processing time, with Finishing B enabling a more aggressive removal rate, making it effective for optimal surface refinement in Nd-Fe-B samples. Thus, applying the electropolishing process after wire electrical discharge machining is a potential process to be applied to manufacturing of Nd-Fe-B magnets.

E. C. Costa (✉)
Department of Mobility Engineering, Manufacturing Engineering Laboratory, Federal University of Santa Catarina, Joinville, SC, Brazil
e-mail: erick.cardoso.costa@ufsc.br

G. Sorgetz · N. S. Heinz · B. C. Santos · F. A. Xavier
Department of Mechanical Engineering, Precision Engineering Laboratory, Federal University of Santa Catarina, Florianópolis, SC, Brazil

M. C. Gomes et al. (eds.), *Innovations in Brazilian Machining*, Lecture Notes in Mechanical Engineering, https://doi.org/10.1007/978-3-031-92819-2_18

Keywords Surface quality · WEDM · Electropolishing · Permanent magnet

1 Introduction

At the moment, the Nd-Fe-B has remained as the most outstanding permanent magnet available, since it presents excellent magnetic properties mainly the high energy stored per volume ($(BH)_{max}$), remanence (B_r) and coercivity (H_{cj}) [1, 2]. These features make the Nd-Fe-B magnet a critical component in various modern technologies related to energy generation and conversion, such as hybrid and electric motors, wind turbine generators, biomedical devices, micro-electro-mechanical systems (MEMS), and other emerging technological applications [3]. Despite the increasing demand for these magnets, their poor mechanical properties pose challenges, leading to reduced performance, shorter lifespan, and a higher susceptibility to mechanical failure during manufacturing [4].

In general, Nd-Fe-B magnets are typically produced using a powder metallurgy process, which involves powder milling, compaction, and magnetic alignment, followed by sintering and post-sintering heat treatments, as described by Cui et al. [5]. Due to shrinkage in volume of the sintered magnet, machining processes are required to ensure stringent design requirements such as geometrical and dimension tolerances, as well as the finished surface for receiving a coating [6]. Machining processes such as slicing, grinding, lapping and polishing are commonly employed in magnet production to ensure the required surface quality. However, the Nd-Fe-B magnet is a typical brittle-hard material, resulting in a difficult-to-machine material, since it is prone to cracking during the machining processes due to its high hardness and brittleness, and low fracture toughness [7, 8].

In this sense, different machining processes have been evaluated to cut Nd-Fe-B magnets for research purposes. Li et al. [9] investigated the surface characteristics of Nd-Fe-B magnets machined using ultrasonic-aided EDM under different cutting conditions. It was observed that the subsurface of the Nd-Fe-B presented a larger thermal damage layer. X-RD analysis revealed that the machined surface was formed mainly by amorphous phases, with some peaks related to diffusion of Cu from the electrode to the magnet. Kruusing et al. [10] also used laser-cutting and EDM processes to obtain plates of Nd-Fe-B. The machining was carried out with the workpiece submerged in deionized water and in an ambient atmosphere. The EDM process provides better accuracy and surface quality when compared with laser-cutting. However, its feed rate and machining time were lower than those obtained with laser-cutting.

Kim et al. [11] studied the effect of diamond cutting to obtain small Nd-Fe-B magnets and evaluated its magnetic properties. It was observed a kink on the demagnetizing curve, reducing both $(BH)_{max}$ and H_{cj}. That result can be attributed to machined surface damage, such as oxidation and deformation. In order to remove the damage, the machined magnets were submitted to an etching process, but the observed kink in the hysteresis curves was not removed. Heinz et al. [12] analyzed the effect of diamond wire sawing and electropolishing processes on the surface integrity and magnetic properties of the Nd-Fe-B. The authors observed microcutting, pittings and craters over the machined surface when diamond wire sawing, which reflected on surface roughness. The phase analysis via X-RD technique revealed some peaks related to the oxides phases on the machined surfaces. For the magnetic property, it was possible to observe that the H_{cj} presented closed values when varying the diamond wire.

Based on the aforementioned, the surface quality of the Nd-Fe-B magnet has a significant effect on the magnetic properties. In this sense, the study aims to provide insight on the feasibility for applying the electropolishing after the wire electrical discharge machining (WEDM) of Nd-Fe-B plates. Different finishing conditions for WEDM were evaluated and the electropolishing was applied under varying current density and time. The surface quality was evaluated in terms of surface morphology, and the surface roughness R_a and R_q parameters were measured. The material removal rate for the electropolishing process was also estimated and the results were analyzed.

2 Materials and Methods

2.1 *Experimental Setup*

The workpiece specimen was a commercial sintered Nd-Fe-B class N42H, which is an anisotropic permanent magnet with a high amount of energy stored per volume, having a BH_{max} of 335 kJ/m^3, and a bulk density of 7.48 ± 0.02 g/cm^3. The workpiece was cut to dimensions of (8.00 × 8.00 × 1.25) mm^3 using a wire electrical discharge machining AgieCut Challenge FW1U that employed a reciprocating coil movement. The tool electrode was an uncoated molybdenum wire with $d_w = 180$ μm and 10% ECOSINT with deionized water (H_2O-De) as dielectric. Based on the crystallographic texture (due to magnetic anisotropy) of the sintered Nd-Fe-B magnet used in this study, the samples were WED-machined following the axis of easy magnetization. The WED-machining conditions are shown in Table 1.

The WED-machined samples were subsequently subjected to an electropolishing (EP) process. The procedure utilized a beaker containing an electrolyte composed of 2% nitric acid in denatured alcohol, with a tungsten wire serving as the electrode. The electropolishing was performed at room temperature, approximately 24 °C, using an

Table 1 WED-machining condition

Conditions	ON (μs)	OFF (μs)	Peak current (A)
Finishing A	31	31	2
Finishing B	31	1	2.5

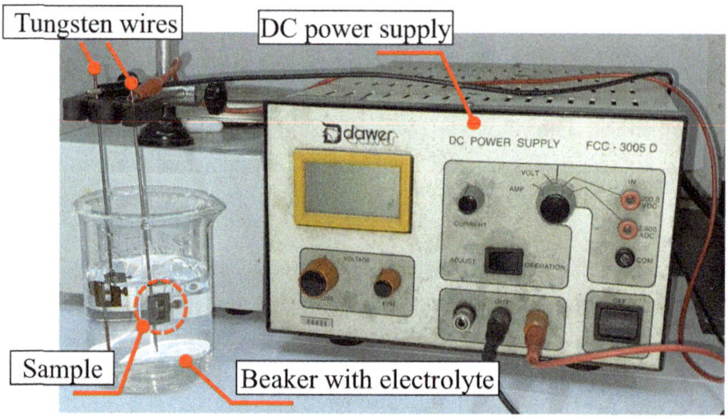

Fig. 1 Electropolishing workbench

Table 2 Experimental electropolishing conditions

Conditions	Current density (A/cm^2)	Time (min)
1	0.3	4
2	0.6	7
3	1.0	10

FCC Dawes model 3005D power supply. The electropolishing setup used in this study is shown in Fig. 1.

The electropolishing experiments were carried out by varying the maximum current density and processing time. A full factorial design of experiments was implemented, leading to a total of 27 experimental runs. The specific conditions for these experiments are summarized in Table 2.

2.2 *Experimental Setup*

To determine the weight of each sample, a METTLER Toledo XS205 DualRange analytical balance was used. This instrument offers a resolution of 0.01 mg for smaller samples and 0.1 mg for larger ones. Weight measurements were recorded both before and after the electropolishing process, allowing for the calculation of the material removal rate (MRR) by assessing the mass difference between the pre- and post-electropolishing states of the samples, taking the material's density into account.

The surface morphology of the samples after WED-machining and electropolishing was characterized using a Hitachi TM-3030 scanning electron microscope (SEM). This SEM analysis was performed under different cutting conditions to observe and document the microstructural features and surface characteristics of the treated samples. The high-resolution imaging provided by the SEM allowed for detailed visualization of surface textures, and the overall morphology resulting from the machining and electropolishing processes.

Surface roughness was conducted using a Taylor Hobson FTS i-120 contact profilometer, adhering to the ISO 21920-3:2021 standard [13] with a cut-off wavelength (λ_c) set at 0.8 mm. The surface profile of each sample was measured three times to ensure accuracy and repeatability of the results. The raw profilometer data was then processed using the MountainsMap Universal software, which complies with ISO 21920-2:2021 standard [14], to calculate the surface roughness parameters. The primary roughness parameters evaluated included the arithmetical mean deviation of the profile (R_a) and the root mean square of the profile (R_q). The reported values represent the arithmetic mean of the measurements, along with their respective standard deviations, providing a comprehensive assessment of the surface finish quality of the samples.

3 Results and Discussion

3.1 Surface Morphologies

Figure 2 shows surface morphologies of the Nd-Fe-B magnet after WED-machining under different conditions, according to the magnitudes shown in Table 1. In general, it can be observed that both micrographs reveal recast material and thermal cracks, which are typical result found for machined surfaces obtained in the electrical discharge machining processes [15]. For the condition of Finishing A (see Fig. 2a), the WEDMed surface looks smoother when compared to that of Finishing B, since it contains some prominent recast layers and surface irregularities, as shown in Fig. 2b. These findings also support the WEDM conditions where the lower peak current, with more off time in Finishing A, most likely caused the impact with less energy discharge and, in turn, reduced the melting of the material and resolidification on the Nd-Fe-B workpiece.

Figure 3 shows the surface morphologies of Nd-Fe-B samples after the electropolishing process, performed under varying current densities (J) and times (t) for samples prepared under the Finishing A condition, as outlined in Table 2. Overall, the electropolishing process enhanced the surface quality by reducing the amount of recast material observed. When analyzing the effect of time, it was noted that prolonged exposure resulted in more pronounced etching on the surface, characterized by cracks, oxide sites, grains, and grain boundaries. Specifically, for a current density of $J = 0.3$ A/cm^2, increasing the processing time from 4 to 10 min generated thermal cracks induced by the WED-machining. However, when maintaining the processing time and increasing the current density from 0.3 to 1.0 A/cm^2, the surface morphology showed significant improvement, with the thermal cracks being not observed and the grain boundaries and individual grains becoming more distinct.

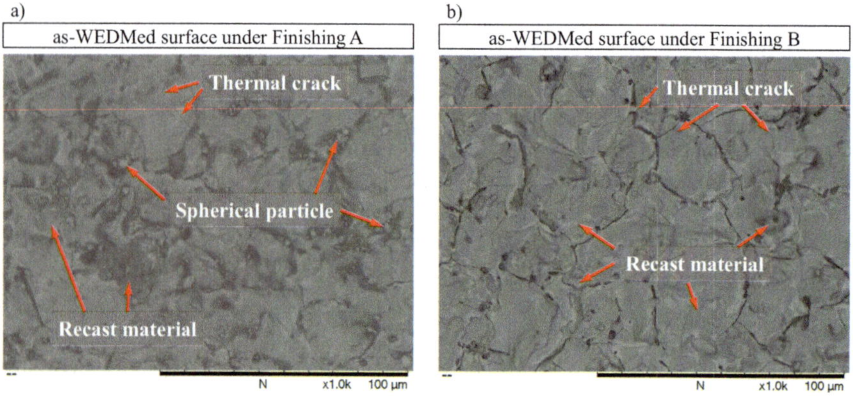

Fig. 2 Surface morphologies of the WEDMed surfaces

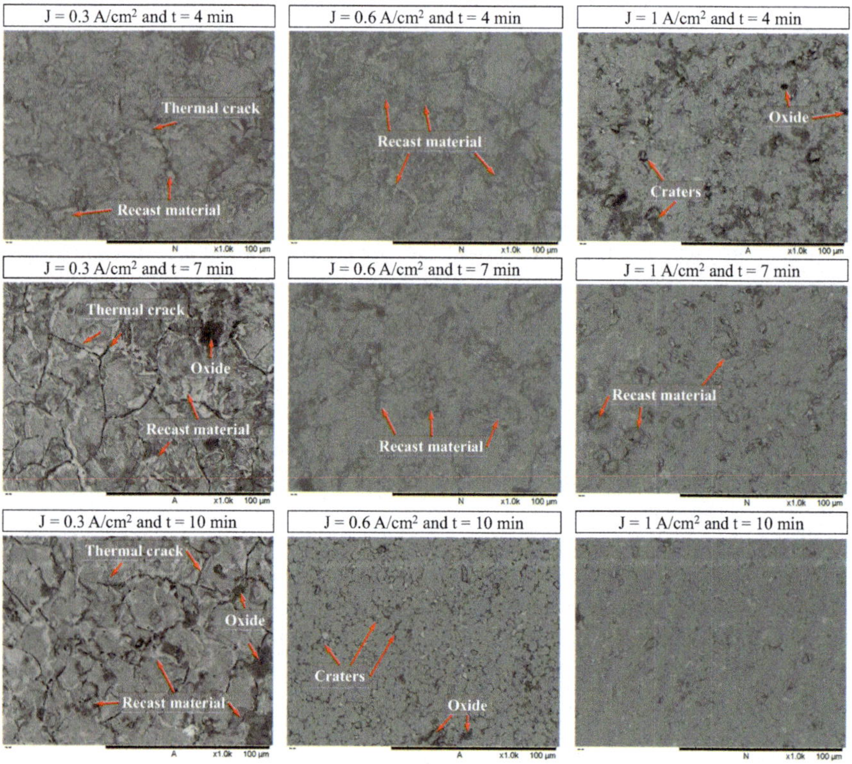

Fig. 3 Surface morphologies after electropolishing for samples obtained under Finishing A

For the condition of WEDM with Finishing B, similar surface morphologies were observed for electropolished Nd-Fe-B samples varying current density and time. Based on the result found for each electropolishing condition applied for both conditions of Finishing A and B, the higher current density used ($J = 1$ A/cm^2) and the longer the electropolishing time ($t = 10$ min), the more uniform and smoother the surface becomes.

3.2 *Surface Roughness*

Figure 4 presents the surface roughness parameters R_a and R_q for both WED-machined (WEDMed) and electropolished (EP) surfaces of the Nd-Fe-B magnet. For reference, the WEDMed surface under Finishing A exhibited $R_a = 2.85 \pm 0.14$ μm and $R_q = 4.27 \pm 0.14$ μm, while Finishing B resulted in $R_a = 1.88 \pm 0.09$ μm and $R_q = 2.67 \pm 0.39$ μm. These values are typical for surfaces machined by electrical discharge machining and correspond well with the surface morphologies observed under different WEDM conditions, as shown in Fig. 2. Notably, applying Finishing B led to a reduction in surface roughness (R_a and R_q) by approximately 64%.

For the electropolished samples, R_a values ranged between 0.50 and 2.30 μm, and R_q values between 0.75 and 3.25 μm for Finishing A, while Finishing B resulted in R_a values between 0.50 and 2.05 μm, and R_q values between 0.50 and 3.00 μm. Comparing Fig. 4a, b, which correspond to Finishing A and B, respectively, it is evident that both R_a and R_q values decreased for the WEDMed samples after electropolishing. The lowest surface roughness values ($R_a = 0.474$ μm and $R_q = 0.607$ μm) were achieved under the Finishing B condition (Fig. 4b). This improvement can be attributed to the reduced amount of recast material on the machined surface obtained under this WEDM condition, which enhanced the effectiveness of the electropolishing process.

The data indicates a clear trend, in which the electropolishing time increasing from 4 to 10 min induces reduction in surface roughness R_a and R_q values. It can be noted that for all cases there was a slight increase of surface roughness for the time condition of 7 min. A significant reduction of surface roughness was obtained when increasing the current density (J) from 0.3 to 1.0 A/cm^2, as can be observed in plots presented in Fig. 4. The most notable improvements in surface finish were achieved at the higher current density of 1.0 A/cm^2 and time of 10 min, where the R_a and R_q values reached their lowest points, demonstrating the effectiveness of the electropolishing process in refining the surface quality of the Nd-Fe-B magnet obtained with WED-Machining.

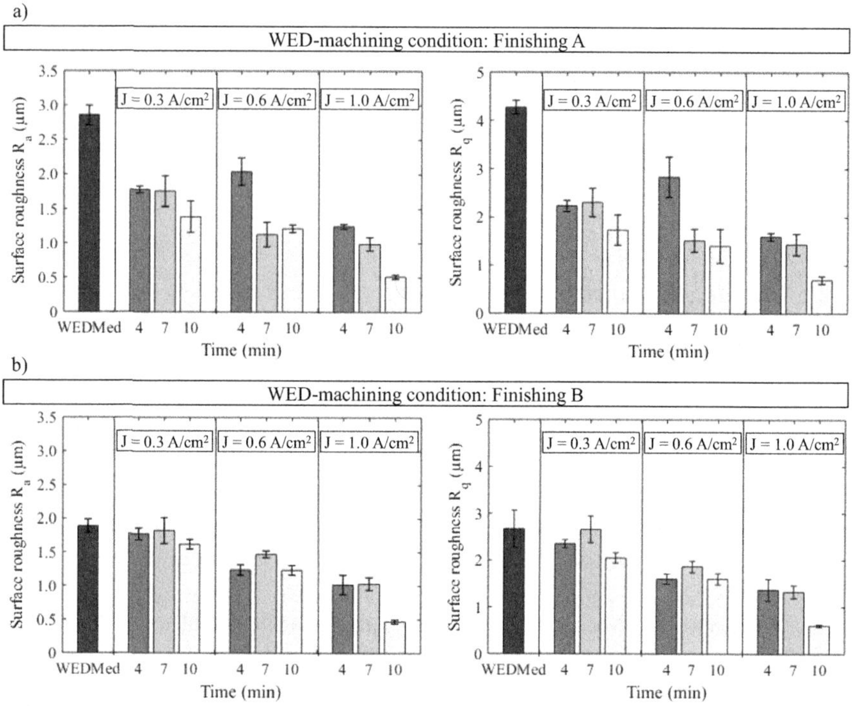

Fig. 4 Surface roughness of the WEDMed and electropolished surfaces obtained under different conditions

3.3 *Material Removal Rate*

Figure 5 shows the material removal under varying current densities and electropolishing time, providing insights into the process for two different WEDM finishing conditions, referred to as Finishing A and Finishing B. As shown in Fig. 5, the MRR increases significantly with both current density and electropolishing time. For the case of the Nd-Fe-B samples obtained under Finishing A, the MRR values kept up to 250×10^5 μm^3/min, while around 300×10^5 μm^3/min for Nd-Fe-B samples WEDMed on Finishing B. Specifically, increasing the current density from 0.3 to 1.0 A/cm^2 led to a significant increase in MRR, with the highest removal rates observed at 1.0 A/cm^2 after 10 min of processing. This trend indicates that higher current densities enhance material removal efficiency during electropolishing.

Additionally, extending the electropolishing time from 4 to 10 min further amplified the MRR, indicating that prolonged electrochemical activity on the sample surface leads to more substantial material removal. Notably, Finishing B consistently exhibited higher MRR values than Finishing A, underscoring the influence of initial surface conditions on the effectiveness of the electropolishing process.

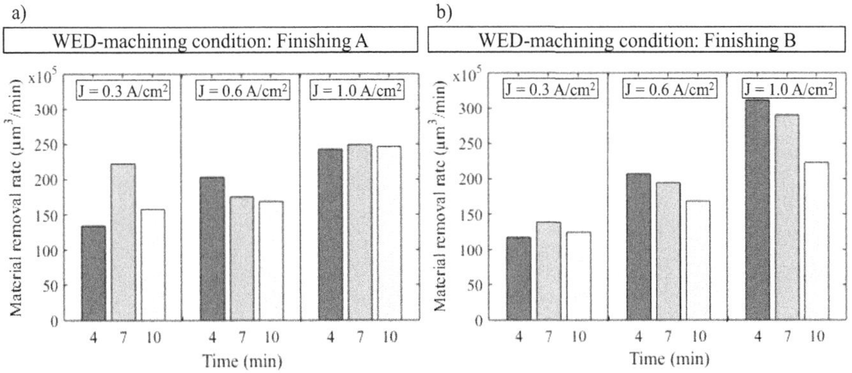

Fig. 5 Material removal rate under different WED-machining and electropolishing conditions

The superior MRR observed in Finishing B can be attributed to surface characteristics produced by the WED-machining process, which favor more effective material removal during subsequent electropolishing. In this sense, it can be stated the MRR during electropolishing is significantly influenced by both current density and processing time, with higher values of each parameter resulting in greater material removal. The condition of Finishing B, in particular, facilitates a more aggressive removal rate, highlighting its potential for achieving optimal surface refinement in Nd-Fe-B samples. These findings underscore the importance of carefully selecting electropolishing parameters to maximize both efficiency and surface quality.

4 Conclusions

This study aims to provide insights on the feasibility for applying the electropolishing process after the WED-machining of the Nd-Fe-B samples. Based on the obtained results, the following conclusions are summarized below:

i. The surface morphologies of the Nd-Fe-B magnet after WED-machining present recast and resolidified material for both WEDM conditions. The EP improved the surface morphologies, since less recast material was observed. The more uniform and smoother the surface was obtained under $J = 1$ A/cm^2 and $t = 10$ min;
ii. The WEDMed surface under Finishing A presented $R_a = 2.85 \pm 0.14$ μm and $R_q = 4.27 \pm 0.14$ μm, while Finishing B resulted in $R_a = 1.88 \pm 0.09$ μm and $R_q = 2.67 \pm 0.39$ μm. The EP decreased surface roughness, where the lowest surface roughness of $R_a = 0.474$ μm and $R_q = 0.607$ μm were found under the condition of Finishing B;

iii. It was observed that the MRR increases significantly with both current density and electropolishing time. Increasing the current density from 0.3 to 1.0 A/cm^2 led to a significant increase in MRR, with the highest removal rates observed at 1.0 A/cm^2 after 10 min. The Finishing B condition facilitated a higher MRR, highlighting its potential for achieving optimal surface quality in Nd-Fe-B.

References

1. Sepehri-Amin H, Hirosawa S, Hono K (2018) Chapter 4—Advances in Nd-Fe-B based permanent magnets. In: Handbook of magnetic materials, 1st edn. Elsevier B.V., pp 269–372
2. Coey JMD (2020) Perspective and prospects for rare earth permanent magnets. Engineering 6(2):119–131
3. Mohapatra J, Liu JP (2018) Chapter 1—Rare-earth-free permanent magnets: the past and future. In: Handbook of magnetic materials, 1st edn. Elsevier B.V., Arlington, Texas, United Sates, pp 1–57
4. Costa EC, Lima AO, Santos MNL, Souza RM, Wendhausen PAP, Xavier FA (2024) Exploring the influence of anisotropy and microstructure of the sintered Nd-Fe-B magnet using nanoindentation and scratching tests. Mater Today Commun 39:1–10
5. Cui J, Ormerod J, Parker D, Ott R, Palasyuk A, Mccall S, Paranthaman MP, Kesler MS, McGuire MA, Nlebedim IC, Pan C, Lograsso T (2022) Manufacturing processes for permanent magnets: Part I—Sintering and casting. J Miner Metals Mater Soc 74:1279–1295
6. Croat JJ (2018) Rapidly solidified neodymium-iron-boron permanent magnets, 1st edn. Elsevier, United Kingdom
7. Costa EC, Lima AO, Santos MNL, Souza RM, Wendhausen PAP, Xavier FA (2023) On the mechanical behavior of the sintered Nd-Fe-B permanent magnet during diamond scratching. J Manuf Process 94:79–93
8. Liu W, Wu J (2008) Mechanical properties and fracture mechanism study of sintered Nd-Fe-B alloy. J Alloys Compd 458:292–296
9. Li L, Wei XT, Li ZY, Cheng X (2014) Corrosion resistance analysis of sintered NdFeB magnets using ultrasonic-aided EDM method. J Mater Eng Perform 1–8
10. Kruusing A, Leppävuori S, Uusimäki A, Petrêtis B, Makarova O (1999) Micromachining of magnetic materials. Sens Actuators, A 74:45–51
11. Kim HT, Kim YB, Kapustin GA (2004) Effects of machining on the magnetic properties of sintered NdFeB magnets. Phys Status Solidi (a) 201:1913–1916
12. Heinz NS, Costa EC, Xavier FA (2023) Manufacturing of micro-scale Nd-Fe-magnets by diamond wire sawing and electropolishing processes. In: 27th ABCM International congress of mechanical engineering. ABCM, Rio de Janeiro, Brazil, pp 1–9
13. ISO: International Organization for Standardization (2021) Geometrical product specifications (GPS)—surface texture: profile method—terms, definitions and surface texture parameters (ISO 21920-3:2021). ISO, Geneva
14. ISO: International Organization for Standardization (2021) Geometrical product specifications (GPS)—surface texture: profile method—rules and procedures for the assessment of surface texture (ISO 21920-2:2021). ISO, Geneva
15. Klocke F (2008) Manufacturing process 2: grinding, honing, lapping. RWTH Springer, Berlin, Germany

Scratch Test Adapted to a Machining Center

Ricardo Augusto Gonçalves, Alexandre Mendes Abrão, Marcelo de Araújo Câmara, Carlos Eiji Hirata Ventura, and Witor Wolf

Abstract The surface behavior of engineering materials is crucial in many areas of modern technology, making it essential to understand the properties and characteristics of these surfaces. One important property is scratch hardness, which can be evaluated through scratch tests. This study explores a novel scratch-testing method developed in a machining center, utilizing an inclined stylus path. AISI 304 austenitic stainless steel samples were used as test specimens and subjected to different processing techniques (milling and deep rolling). A single-tip diamond dresser served as stylus. The results indicated that this new technique is promising, as the surface roughness and hardness variations among the tested samples affected the scratch hardness and friction coefficient outcomes. Additionally, this testing method may provide an alternative approach for investigating the minimum chip thickness.

Keywords Scratch hardness · Friction coefficient · AISI 304 austenitic stainless steel · Deep rolling · Milling

1 Introduction

For most engineering applications, it is essential to identify the characteristics and properties of a part or component's surface, especially if this surface is in contact or in relative movement with another surface. These characteristics, such as hardness, resistance to abrasive wear, scratch hardness, coatings adhesion, fatigue and corrosion resistance, and lubrication capability, can determine how the materials in contact

R. A. Gonçalves (✉)
Federal University of Jequitinhonha and Mucuri Valleys, Diamantina, Minas Gerais, Brazil
e-mail: ricardo.augusto@ict.ufvjm.edu.br

A. M. Abrão · M. de A. Câmara
Universidade Federal de Minas Gerais, Belo Horizonte, Minas Gerais, Brazil

C. E. H. Ventura · W. Wolf
Federal University of São Carlos, São Carlos, São Paulo, Brazil

M. C. Gomes et al. (eds.), *Innovations in Brazilian Machining*, Lecture Notes in Mechanical Engineering, https://doi.org/10.1007/978-3-031-92819-2_19

will behave. The scratch test is one way to characterize surfaces regarding scratch hardness.

The scratch test consists of scratching the surface of a sample and knowing the applied load, the scratching speed, and the geometric characteristics of the stylus tip. The results are obtained based on the scratch morphology and the normal and tangential components of the force applied to scratch the surface [1]. The scratch test is applied to several sectors, including the automotive, biomedical, cutting tools (hard coatings), decorative coatings, electronics, polymeric materials, and metallic and advanced ceramics industries [2]. A scratch test is capable of [1, 3]:

- determining the forces acting on the different phases of the same material;
- evaluating the efficiency of a lubricant;
- analyzing a microstructural effect on the life of a component;
- obtaining scratch hardness, coefficient of friction, and specific energy of materials;
- characterizing the mechanisms of deformation and material removal;
- evaluating or classifying materials about abrasion resistance;
- evaluating the adhesion between the coating and the substrate.

The scratch test, or linear sclerometer, is standardized by ASTM G171-03—Standard Test Method for Scratch Hardness of Materials Using a Diamond Stylus [4]. This standard describes the laboratory procedures for determining the scratch hardness of the surface of solid materials using a conical diamond stylus with a hemispherical tip. Some test variations include the low-speed, low-speed in situ, and high-speed tests. In the low-speed test, the scratching element usually moves along the surface at a speed of a few millimeters per second, thus generating a single-pass groove. The element's configuration can also generate a multiple-pass groove or slide in circular paths. It is important to note that this standardized test is performed by applying a constant load to the stylus. The in situ low-speed test uses equipment that performs sclerometer tests inside a scanning electron microscope (SEM), which is used in more detailed studies of wear mechanisms and groove formation. Finally, the high-speed test can be performed in two ways, with the scratching element mounted on a pendulum or the periphery of a rotating disk [4].

Alternatively, the use of different equipment to perform the scratch test has also been reported [5]. In this study, a Rockwell C indenter mounted on the spindle of a CNC vertical machining center performed scratch tests at constant load in order to relate different cutting conditions in the milling of AISI D2 steel (substrate) to the performance of the AlCrN coating in terms of adhesion. It is important to note that almost all scratch tests are performed with a linear displacement and at a constant depth.

Thus, the main goal of this study is to investigate a new approach for performing scratch tests using a machining center and a single-tip diamond dresser as a stylus developing a ramp path. Additionally, the tribological behavior of AISI 304 austenitic stainless steel surfaces processed by milling and milling followed by deep rolling is analyzed.

2 Materials and Methods

The material used as the test specimen is AISI 304 austenitic stainless steel. After machining, samples with dimensions of 30 × 30 × 12 mm were solution annealed (maximum temperature of 1200 °C) to inhibit carbide formation at the grain boundaries, thus ensuring austenite formation.

The samples were mounted in a clamping device, see Fig. 1a, and then in the dynamometer. Figure 1b shows the stylus tip (diamond dresser). The normal and tangential components of the scratching force (Fig. 1c) were measured using a Kistler 9272 piezoelectric dynamometer. The signal acquired by the dynamometer was sent to a Kistler 5073 amplifier, then to the National Instruments 6366 acquisition board, and, finally, was recorded by the Signal Express software.

After the scratching tests, the coefficient of friction (μ) was estimated by the ratio of the tangential force (F_T) or friction force to the normal force (F_N) [6], see Eq. (1). ASTM G171-03 standard names this coefficient after stylus drag coefficient (D_{scr}), analogously calculated by dividing the scratching force (F_{scr}) by the normal force (F_N), according to Eq. (2).

$$\mu = \frac{F_T}{F_N} \tag{1}$$

$$D_{scr} = \frac{F_{scr}}{F_N} \tag{2}$$

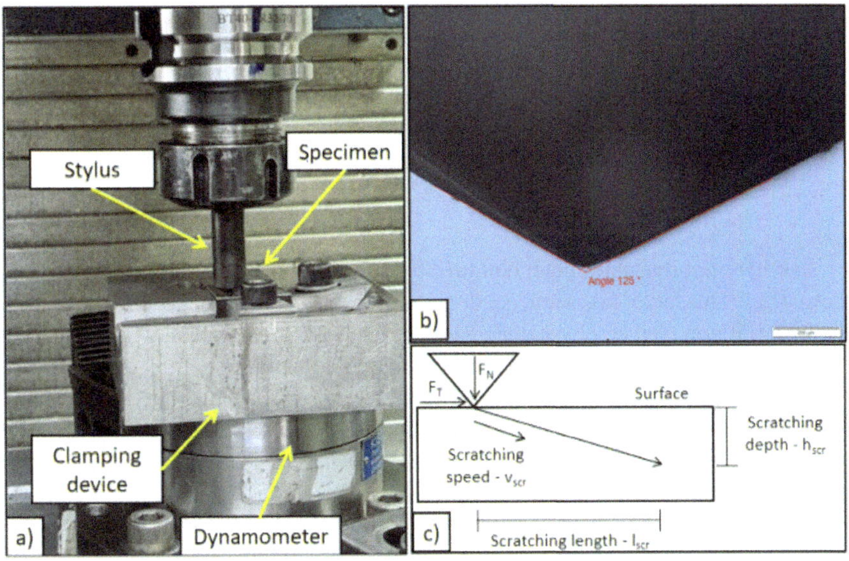

Fig. 1 **a** Stylus, test specimen and clamping device, **b** stylus tip and **c** scratch test input parameters

The preparation of the samples before the scratch tests consisted of face milling alone and face milling followed by deep rolling. The samples were initially face milled with a cutting speed of 200 m/min, feed per tooth of 0.1 mm/rev, and axial depth of cut of 0.1 mm. Afterwards, one sample was deep rolled at a speed of 100 mm/min with a pressure of 160 bar and applying five rolling passes.

The average hardness of the samples was obtained from ten measurements performed on a Buehler 1600–6300 hardness tester using a load of 200 g. The scratches were qualitatively analyzed using an Olympus CX31 optical microscope. Then, an Alicona Infinite Focus SL confocal microscope was used to obtain the scratches profile) and measure the surface roughness of the samples.

The scratch tests were performed on a vertical machining center utilizing a single diamond dresser (Fig. 1b) as the stylus and ground to a tip angle of 125°. During the scratch tests, the stylus developed a translational movement in the X and Z directions, thus resulting in a scratching speed (v_{scr}) of 2 mm/min, scratching depth (h_{scr}) of 25 μm and scratching length (l_{scr}) of 2.5 mm. Three profiles (cross-sections) from each scratch were obtained at three different positions (beginning, middle, and end of the scratch) in order to calculate the area of displaced material (pile-up) and material removed (groove) during the tests. The surface roughness of the samples was measured obtaining average values and standard deviations of 200 profiles. The surface roughness area parameters were also measured in a 1 mm × 1 mm region.

The scratch hardness represents the resistance of a solid surface against penetration by a moving stylus of a given tip radius under a constant normal force and speed. It's calculated through the ratio of normal force applied by the stylus and the projected scratch contact area, assuming that the hemispherical tip stylus produces a groove whose main surface has a curvature radius r, the stylus tip's radius. Therefore, the projected contact surface area is a semicircle which diameter is the final scratch width [3]. Thus, the scratch hardness (H_S) is obtained according to Eq. (3), where w is the scratch width. The maximum normal force measured in each test was used in the calculations.

$$H_S = \frac{8 \times F_N}{\pi \times w^2} \tag{3}$$

Another important parameter obtained from the scratch test is the material removal factor (f_{ab}). This factor takes into account the areas of displaced material (pile-up) and the material removed (groove) during the scratch test. It is calculated by Eq. (4) and can range from 0 (microploughing or pure material displacement) to 1 (pure micro-cutting) [7]. The areas of interest in the scratch section were calculated according to [8]. The areas were obtained using Image J software to generate approximate polygons in the regions of displaced and removed material. Three profiles (beginning, middle, and end) were analyzed for each scratch. Figure 2a–c shows the methodology for obtaining the scratch cross-section areas for the stylus entry into the deep rolled sample.

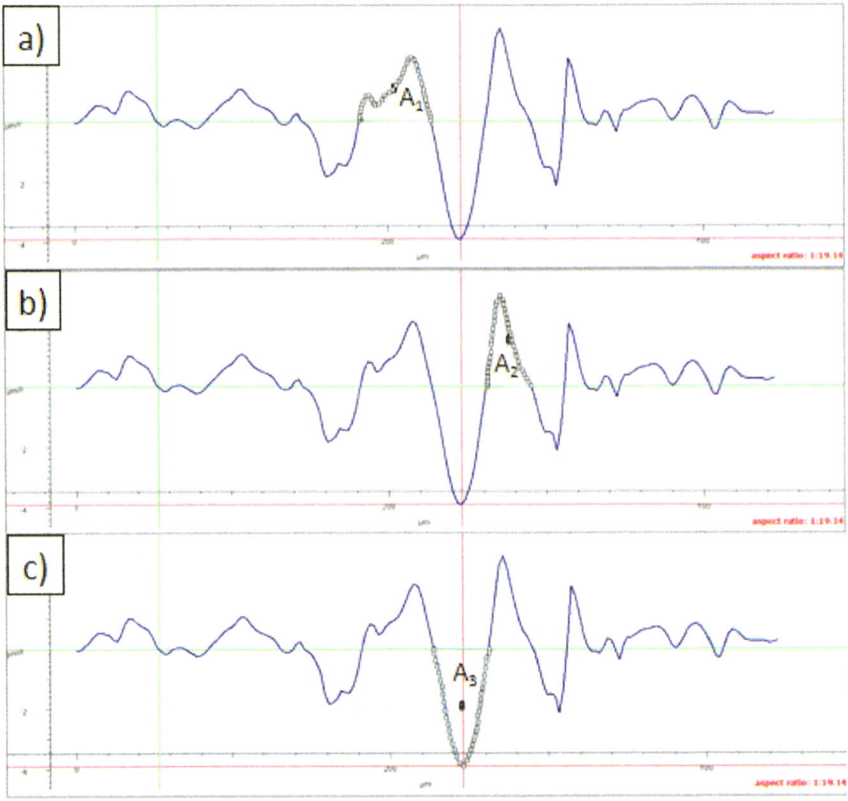

Fig. 2 Scratch cross-section areas **a** A_1 (pile-up), **b** A_2 (pile-up) and **c** A_3 (groove)

$$f_{ab} = \frac{A_3 - (A_1 + A_2)}{A_3} \quad (4)$$

3 Results

The hardness test indicated an average value of 291.37 HV (standard deviation of 10.61 HV) for the milled sample and 383.74 HV (standard deviation of 14.05 HV) for the milled and deep rolled sample. Deep rolling induces work hardening, which significantly contributes to the increase in hardness on the surface and subsurface of the material [9].

Figure 3 shows the surfaces of the milled and deep rolled samples. In a visual analysis, it is possible to observe tool feed marks on the milled sample, which are not evident on the deep rolled surface. This smoother surface is due to the characteristics of the process in which the tool compresses the surface, plastically deforming the roughness peaks generated during the previous machining operation, thus improving the surface quality of the machined parts [10].

The measured surface roughness parameters are presented in Table 1. Regarding the surface finish, the deep rolled sample presented superior quality compared with the milled sample. After deep rolling, the arithmetic mean deviation (R_a) decreased from 0.29 to 0.20 μm. Likewise, the total height of the profile between the deepest valley and the highest peak in the evaluation length (R_t) decreased from 2.01 to 1.47 μm. It is important to highlight the values obtained from the standard deviation (S_d) of the measured parameters R_a and R_t, indicating a statistically significant difference between the averages. These results corroborate the results obtained by [11]. According to the authors, the topography of the surface resulting from deep rolling is characterized by flat peaks and valleys filled with material due to plastic deformation, which may justify the decrease in these roughness parameters. Furthermore, the reduced height of the peaks and valleys (R_{pk} and R_{vk}) observed on the deep rolled surface were smaller than on the milled surface. In agreement with the analysis performed, the reduction of R_{pk} is more significant than that observed for R_{vk}. Lower values of R_{vk} represent a reduction in the coefficient of friction [12]. Finally, it is important to highlight that the area surface roughness parameters (S_a and S_q) also decreased after deep rolling.

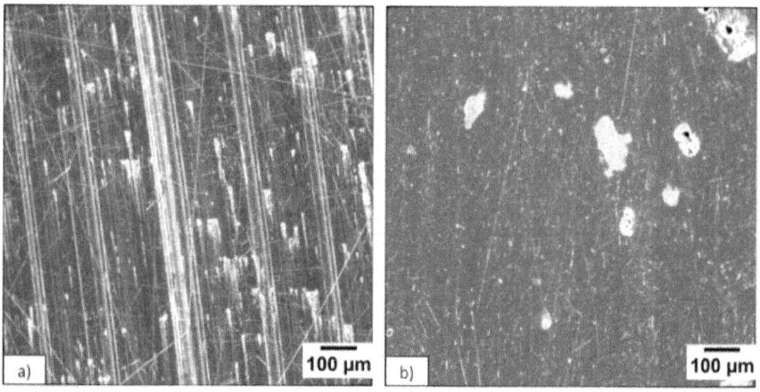

Fig. 3 Surface's images of: **a** milled and **b** deep rolled AISI 304 austenitic stainless steel samples

Table 1 Surface roughness parameters of milled and deep-rolled AISI 304 austenitic stainless steel samples

Surface roughness (μm)	Milled	Deep rolled
R_a	0.29 ($S_d = 0.01$)	0.20 ($S_d = 0.01$)
R_t	2.01 ($S_d = 0.14$)	1.47 ($S_d = 0.37$)
R_{pk}	0.48	0.08
R_{vk}	0.20	0.16
S_a	0.21	0.13
S_q	0.29	0.18

Figure 4 shows the scratches produced in the milled and deep rolled samples. The wider width along the entire scratch is evident in the milled samples, which presented lower hardness than the deep rolled samples. For the milled samples, the scratch widths in the evaluated section were 360 and 310 μm, while for the deep rolled samples they were 186 μm and 187 μm. It is also possible to observe a more significant displacement of material at the edges of the scratches of the milled samples. However, this behavior will be discussed in more detail together with the material removal factor.

Figure 5 show the behavior of the normal and tangential forces measured during the scratch tests on the milled and deep rolled AISI 304 austenitic stainless steel samples. It is possible to observe the proportionality between the friction force and the normal force, illustrating the first law of friction established by Leonardo Da Vinci in 1493 [6]. The highest values of the maximum normal force obtained in the scratch tests on the milled samples were 102 N (Fig. 5a) and 109 N (Fig. 5b). For the deep rolled samples, thighest normal forces were 52 N (Fig. 5c) and 64 N (Fig. 5d). In this case, the better surface finish of the deep-rolled sample may have led to less resistance to the stylus movement despite the greater hardness of this material compared to the milled sample. It is also possible to change the shear stress of the interface between the surfaces of the stylus and the test specimen due to the work hardening of the rolled sample.

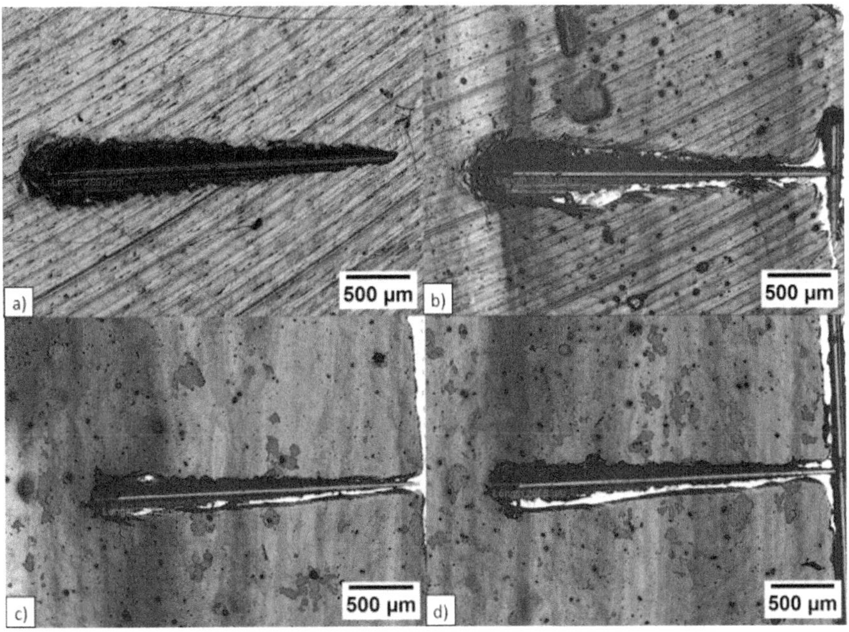

Fig. 4 Scratches produced on: milled (**a**, **b**) and deep rolled (**c**, **d**) AISI 304 austenitic stainless steel samples

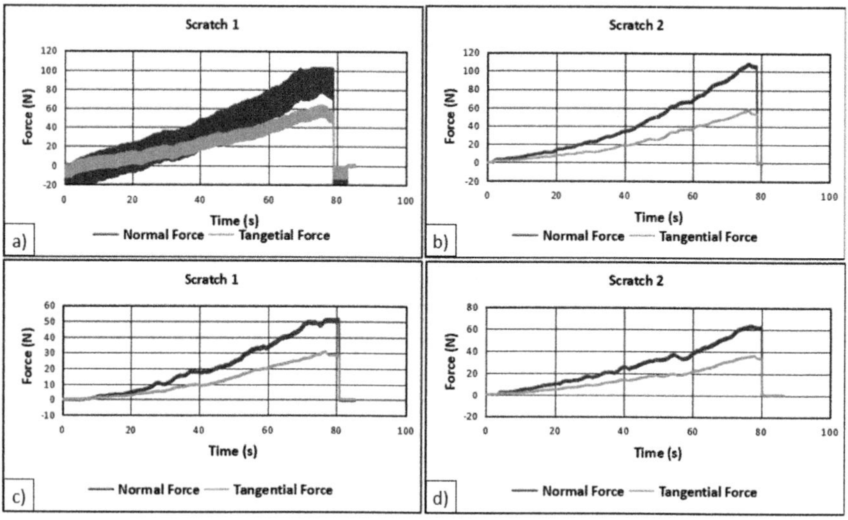

Fig. 5 Normal and tangential forces measured during the scratch test of the milled (**a**, **b**) and deep rolled (**c**, **d**) samples of AISI 304 austenitic stainless steel

Figure 6 shows the results of the coefficient of friction and scratch hardness. On average, the coefficient of friction between the diamond stylus and the deep rolled sample was lower ($\mu = 0.540$) than the milled sample ($\mu = 0.598$), see Fig. 6a. However, since the standard deviation of the measurements overlaps, it is not possible to state that there is a statistical difference between the averages. The friction force deals with the response of the interaction between the asperities of two surfaces. The coefficient of friction, in turn, depends on several factors, such as the surface finish, hardness of the materials in contact, and temperature, among others. In this case, the deep rolled sample's lower surface roughness contributed to the friction coefficient reduction during the tests performed on this sample.

The scratch hardness of the deep rolled steel sample ($H_S = 4258$ MPa) was, on average, higher than that of the milled sample ($H_S = 2454$ MPa), as shown on Fig. 6b. In the calculation of the scratch hardness, the normal force is directly proportional to the hardness, while the scratch width is inversely proportional with a quadratic influence. Thus, the wider width of the scratches obtained in the tests performed on the milled samples was more influential in reducing the scratch hardness, even though the normal force values obtained during the tests were higher for these samples.

Finally, Fig. 7 shows the calculated values of the material removal factor (f_{ab}) at the beginning, middle, and end of the scratches of the milled and deep rolled samples. When analyzing the results obtained at the stylus entrance (beginning of the scratch), some negative values close to zero are observed. According to Eq. (4), these negative values would indicate that the area of displaced material would be larger than that of the groove. This result may be due to irregularities on the surface that influenced the estimate of the area of the top of the displaced material. The

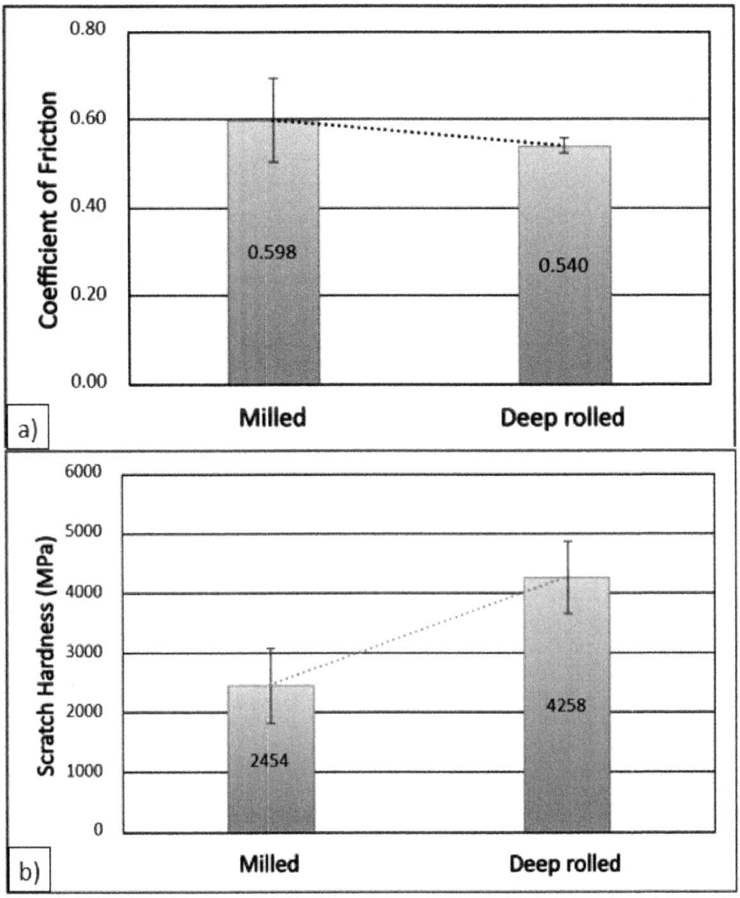

Fig. 6 **a** Coefficient of friction and **b** scratch hardness of milled and deep rolled samples of AISI 304 austenitic stainless steel

values close to zero (except for scratch 1 performed on the milled sample) show that at the beginning of the scratch, that is, at the stylus entrance, the phenomenon of pure microploughing or pure displacement of material prevails due to the small penetration depth or cutting thickness (19.66 μm and 3.39 μm for the milled samples and 3.89 μm and 4.67 μm for the deep rolled samples). On the other hand, in the middle and at the end of the scratches (except for scratch 2 of the milled sample), the values varied between 0.456 (middle of scratch 1 of the milled sample) and 0.763 (end of scratch 2 of the milled sample), indicating the occurrence of two phenomena: microcutting with microploughing (material displacement). It can be concluded that the proposed approach can be used as an alternative in investigating the minimum cutting thickness, as long as several profiles along the scratch are

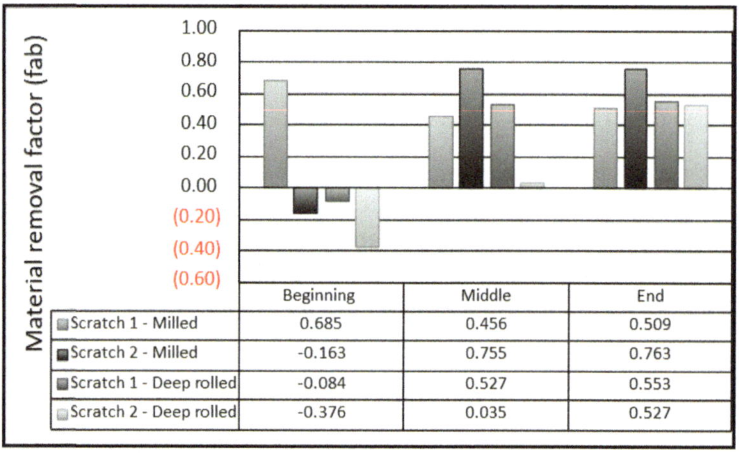

Fig. 7 Material removal factor for the milled and deep rolled samples of AISI 304 austenitic stainless steel

analyzed. The scratching test performed on the different samples showed no tendency for the material removal factor to increase or decrease.

4 Conclusion

The proposed investigation into a new approach to the scratch test developed in a machining center led to relevant conclusions concerning the tribological behavior of milled and deep rolled AISI304 austenitic stainless steel samples:

- Deep rolling resulted in an improvement in the surface finish of the samples as well as an increase in the material hardness;
- The values of normal force and tangential force acquired during the tests provided an estimate of the coefficient of friction between the materials involved in the test. In addition, based on the maximum normal force and the width of the scratch, it was possible to estimate the scratch hardness;
- The sample subjected to deep rolling presented the highest scratch hardness;
- Due to its lower average surface roughness,the coefficient of friction was lower between the diamond tip stylus and the deep rolled sample;
- The material removal factor indicated microploughing at the beginning of the scratches and microploughing with microcutting at the middle and end of the scratches.

References

1. Sousa FJP, Tridapalli D, Pereira M, Flesch CA, e Alarcon OE (2006) Evaluation of measurement uncertainties for a scratching tester. Measurement 39(7):594–604
2. Anton-paar Homepage, https://www.anton-paar.com/corp-en/products/group/scratch-tester/. Last accessed 07 Sep 2024
3. Jacobsson S et al (1992) ASM handbook Volume 18: Friction, lubrication and wear technology: scratch testing. ASM International, Ohio
4. ASTM (2009) G171: Standard test method for scratch hardness of materials using a diamond stylus, 3rd edn. ASTM International, West Conshohocken
5. Carreira AdeS, Almeida G, Haring GM, Ventura CEH (2021) Surface integrity of machined AISI D2 steel and its effect on the adhesion of a PVD-AlCrN coating. Int J Adv Manuf Technol 112(9–10):2705–2715
6. Hutchings I, Shipway P (2017) Tribology: friction and wear of engineering materials, 2nd edn. Butterworth-Heinemann, Oxford
7. Franco A, Sinatora A (2017) Material removal factor (fab): a critical assessment of its role in theoretical and practical approaches to abrasive wear of ductile materials. Wear 382–383:51–61
8. Da Silva BO (2023) Investigação dos mecanismos de remoção de material de insertos de metal duro de diferentes classes durante a retificação. Dissertação, Pós-Graduação em Engenharia Mecânica da Universidade Federal de São Carlos, São Carlos, p 60
9. Klocke F, Bäcker V, Wegner H, Feldhaus B, Baron HU, Hessert R (2009) Influence of process and geometry parameters on the surface layer state after roller burnishing of IN718. Prod Eng 3(4–5):391–399
10. Oliveira DA, Martins AM, Magalhães FdeC, Abrão AM (2022) Characterization of the topography generated by low plasticity burnishing using advanced techniques. Surf Coat Technol 448
11. Oliveira DA, Brito PP, Magalhães FdeC, Azzi PC, Ardisson JD, Abrão AM (2023) Influence of low plasticity burnishing on the formation of strain induced martensite in the surface layer. J Mater Res Technol 27:3722–3743
12. Tekkaya AE et al (2013) Friction analysis of thermally sprayed coatings finished by ball burnishing and grinding. Prod Eng Res Dev 7(6):601–610

Texture Behavior in Ni–Cr–Fe Alloys by Different Lateral Infeed Machining in WEDM Process

Giovani Conrado Carlini, Igor dos Santos Roik, Rodrigo Blödorn, and Fred Lacerda Amorim

Abstract This study takes a unique approach to investigate and explore the texture of the resolidified layer on Ni–Cr–Fe-based materials that form on surfaces manufactured by wire electrical discharge machining (WEDM). The surfaces were machined while keeping the open voltage, current, and other generator parameters constant. Five grids of lateral feed percentage Δ_y were varied, based on the wire diameter d_w (180 μm), acting directly on the change of the front working clearance with variations of $\Delta_y = 100, 75, 50, 25$, and 10%. The results show that the lateral feed affects the roughness of Sa and Sz, respectively. Inconel 718 presented the smallest variations and the largest amplitude Sz. The stainless steel AISI 304 resulted in smaller distortions, mainly at $\Delta_y = 10$ and 25%. In both materials, the application of $\Delta_y =$ 100% resulted in the worst textures due to the decreased washing efficiency in the working gap. These findings have practical implications for the manufacturing of Ni–Cr–Fe-based materials. An Sdr—Smr_2 mapping was performed for the hybrid and functional characterization of the texture behavior, and no correlations were found between the variation of the tested parameters.

Keywords WEDM · Texture surface · Machining · Inconel 718 · Stainless Steel AISI 304

G. C. Carlini (✉)
Instituto Federal de Santa Catarina—IFSC, Jaraguá do Sul, SC, Brazil
e-mail: giovani.carlini@ifsc.edu.br

I. dos Santos Roik · R. Blödorn
Centro Universitário de Brusque—UNIFEBE, Brusque, SC, Brazil

F. L. Amorim
Pontifícia Universidade Católica do Paraná—PUCPR, Curitiba, PR, Brazil

M. C. Gomes et al. (eds.), *Innovations in Brazilian Machining*, Lecture Notes in Mechanical Engineering, https://doi.org/10.1007/978-3-031-92819-2_20

1 Introduction

The study on the texture behavior in Ni–Cr–Fe alloys by different lateral infeed machining in the WEDM process is significant. It addresses the growing need for high-precision features in modern engineering applications, particularly those requiring the production of functional parts from advanced materials with heightened mechanical and thermal properties and high corrosion resistance. These features must adhere to tight tolerances, high accuracy, complex geometries, and micro parts [1].

Among non-conventional machining processes, Wire Electrical Discharge Machining (WEDM) stands out as an essential technique for processing advanced materials, particularly in modern engineering applications requiring high-precision components with superior mechanical. WEDM is recommended for machining various materials, including hardened steels, carbides, ceramics, alloys, and superalloys [1].

This study focuses on Ni–Cr–Fe alloys, specifically Stainless Steel AISI 304 and Inconel 718, due to their enhanced physical, chemical, and technological properties. AISI 304, widely used in the food industry, machinery, and textile manufacturing, is notable for its corrosion resistance, weldability, and ductility, owing to its austenitic structure and non-magnetic nature [2], Conversely, Inconel 718 is a high-performance alloy that retains mechanical properties at elevated temperatures, making it challenging to machine using conventional methods. Its deformation hardening and high heat retention during cutting increase tool wear and hinder material removal, particularly under low feed rates and shallow depths of cut [3].

In WEDM, non-electrical parameters such as dielectric fluid, pulse methods, and kerf control significantly influence outcomes [4]. The literature highlights a gap in evaluating lateral infeed variations, particularly for materials like Inconel 718 and AISI 304. Lateral infeed affects the distribution of discharge energy We, impacting material removal volume, wire feed rate vf, and surface texture.

Recent studies have explored three-dimensional parameters for characterizing surface textures when traditional metrics such as Ra/Sa or Rz/Sz are insufficient to describe a specific application. Pagani et al. [5] described elements to characterize free-form surfaces using three-dimensional parameters such as Sa, Sq, Ssk, Sku, Sdq, and Sdr. They aimed to identify combinations of parameters that could characterize surfaces, forming a topological map. Using a classification matrix and the Spearman correlation technique, Qi et al. [6] identified correlations among amplitude, functional, hybrid, and spatial parameters, enhancing surface characterization techniques.

Carlini et al. [7] analyzed textures of Inconel 718 processed by sink EDM using different electrodes. They found a strong negative correlation ($\rho = -0.91$) between Smr_2 (functional parameter) and Sdr (hybrid parameter). Their findings showed that surfaces with smaller distortions had higher peak material components, influencing load-bearing capabilities.

Buk [8] Conducted experiments on Inconel 718 removal by WEDM involved complete channel cutting (100% wire diameter, d_w) for thinning and lateral infeed (Δ_y) of 30, 50, and 70 μm from $d_w = 250$ μm for finishing. Results showed lower Δ_y reduced roughness and minimized microcracks but did not completely remove resolidified layers from the thinning phase.

This study builds on these findings by examining the surface textures resulting from five lateral infeed levels during WEDM of Ni–Cr–Fe alloys. Maintaining consistent discharge energy, the investigation evaluates surface texture using three-dimensional amplitude parameters (Sa and Sz) and analyzes their correlation with Smr_2 and Sdr. The aim is to provide quantitative insights into how lateral infeed variations influence material behavior and surface quality during WEDM. This contributes to optimizing machining parameters for advanced materials like Inconel 718 and Stainless Steel AISI 304.

2 Experimental Materials and Methods

This study presents a methodology specifically observing the texture resulting from the Wire Electrical Discharge Machining (WEDM) process on two Ni–Cr–Fe alloys when five variations of lateral infeed Δ_y are used. The alloys selected for testing are frequently found in current publications due to their properties and applications in various sectors, including Inconel 718 and Stainless Steel AISI 304. In WEDM operations, the grinding process generally uses a distribution region of 180° in relation to the action of the tool wire (main cut), utilizing the nominal d_w diameter.

For regrinding or semi-finishing (trim cut) operations, lateral infeed Δ_y is used to reduce manufacturing time compared to finishing time, with reduced discharge energy W_e and smaller roughness amplitude values. From this, the five variables associated with Δ_y were stipulated, keeping the electrical quantities constant, and the topographic changes of the removed surfaces were evaluated. Figure 1 shows graphically the main WEDM nomenclatures described in VDI standard 3402 [9].

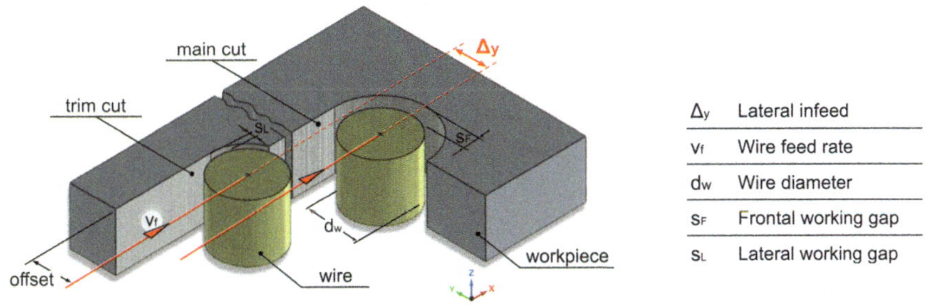

Fig. 1 Lateral infeed (Δ_y) and other WEDM definitions nomenclatures described in the VDI standard 3402 [9]

2.1 Materials

The tests aimed to evaluate the effects of independent electrical input parameters on technological performance in dielectric deionized water and to determine the impact of equal discharge energy parameters on Ni–Cr–Fe materials. Table 1 presents the chemical composition of the Inconel 718 alloy and Stainless Steel AISI 304. Samples with dimensions of 10 mm × 10 mm × 50 mm and two replicates were machined.

As Table 1 shows, the different chemical distributions of the material result from changes in the material properties. The Inconel 718 plays complementary roles in mechanical strength. Niobium (Nb) promotes the formation of γ'' phases (Ni_3Nb), and Aluminum (Al) contributes to the formation of the γ' phase (Ni_3Al), reinforcing the alloy and stabilizing the matrix. Manganese (Mn) improves corrosion resistance and acts in deoxidation [10]. Together, these additions optimize mechanical and thermal properties at high temperatures. Table 2 shows the following mechanical and physical properties that characterize the Ni–Cr–Fe alloys.

Table 2 compares the properties of Inconel 718 and Stainless Steel AISI 304, highlighting notable differences in key characteristics. To achieve effective material removal in the WEDM process, a combination of electrical and non-electrical parameters is essential. The significant impact of electrical parameters on process performance. While it is common to represent a specific WEDM regime in terms of the discharge energy delivered to the working gap, it is crucial not to overlook the influence of material properties on the process [12].

Table 1 Chemical composition (wt%) of the Ni–Cr–Fe alloys. Adapted from Naeim et al. [11]

Ni–Cr–Fe alloys	Element (%)							
	Ni	Cr	Fe	C	Mn	Nb	Al	Mn
Inconel 718	53.0	18.0	21.0	0.08	0.35	4.8	0.30	0.80
Stainless Steel AISI 304	8.0	20.0	67.0	0.08	2.00	*	*	*

*Indicates that the respective element is not present in the alloy

Table 2 Principal mechanical and physical properties of Ni–Cr–Fe alloys

Properties	Ni–Cr–Fe alloys		Unit
	Inconel 718	Stainless Steel AISI 304	
Density	8.2	7.9	g/cm^3
Electrical resistivity	125	70	µΩ·cm
Thermal conductivity	11.2	16.2	W/m·K
Thermal expansion	12.1	17.2	$\times 10^{-6}$/K
Tensile strength	1426	505	MPa
Melting point	1345	1450	°C

Adapted from Carlini et al. [4] and Naeim et al. [11]

Table 3 Independent input variables for the WEDM process

Variable	Symbol	Value	Unit
Lateral infeed	Δ_y	10; 25; 50; 75 and 100	%
Discharge voltage	$\hat{u}_i$	– 70.0	V
Discharge current	$\hat{i}_e$	7.0	A
Discharge duration	t_e	30.0	µs
Interval time	t_o	75.0	µs
Dielectric inlet pressure	P_{in}	0.25	MPa
Wire tension	–	1200	N/mm^2
Wire run-off speed	W_s	8.0	m/s

The machining experiments of WEDM using molybdenum wire on Inconel 718 and AISI 304 were carried out with an isoenergetic generator, maintaining equal independent input variables. The tests were conducted on a GF AgieCharmilles® FW1U CNC 5-axis machine with a wire reuse system using a reciprocating coil. These variables included a discharge current $\hat{i}_e$ of 7 A and the open-circuit voltage $\hat{u}_i$ of – 70 V. Discharge duration and interval time between pulses remained constant for all lateral infeed changes. Thus, all discharge control parameters were kept consistent to understand the direct effect of lateral infeed. Table 3 presents the independent input variables used in the WEDM machine.

2.2 *Methods*

The machining duration and the wire feed rate vf (mm/min) for each operation were controlled using the software and interface available on the equipment WEDM. Five grades were conducted, varying the percentage of lateral infeed Δ_y, based on the diameter of the wire d_w (0.18 mm) and acting directly on changing the frontal working slit with variations of $\Delta_y = 100, 75, 50, 25$, and 10%. Thus, the width of the electrical discharges through the wire in relation to its diameter resulted in $\Delta_y =$ 180 µm, 135 µm, 90 µm, 45 µm and 18 µm, respectively. Figure 2 schematically represents the arrangement of the lateral infeed Δ_y in the front and the top view.

After the machining process, the samples were manually brushed (polyester bristle) and washed were subjected to ultrasonic cleaning at a frequency of 40 kHz for 3 min in C_3H_8O alcohol. Subsequently, the analysis of surface texture commences with a detailed examination of the three-dimensional surface morphology using focus variation microscopy (Bruker Alicona® G5) at an optical lens with a 20x magnification, covering an area of 5.0 mm × 5.0 mm. The Gaussian filter of the standard ISO 16610-71 order 1 and a 250 µm best-suited cut-off sampling length (λc) was established according to ISO 25178–2 were selected. A vertical resolution of 50 nm and a lateral one of 1.5 µm were selected.

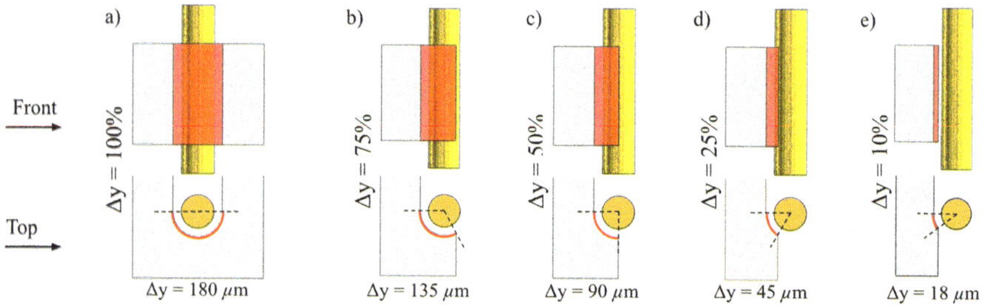

Fig. 2 Graphical representation of area application of electrical discharges when changing the lateral infeed increment (Δ_y)

The Alicona MeasureSuite 5.3.1® software was selected to process parameter data. The use of this software ensures an accurate and comprehensive analysis of the surface texture and relevant hybrid measurements. The amplitude parameters analyzed were Sa and Sz. According to Czifra and Barányi [13], for the functional parameter components, the Firestone-Abbott curve served as the foundational basis, allowing for the collection of relevant data on the rolling area, denoted as Smr_2 (%), as well as the surface distortion percentage, represented by Sdr.

Following the collection of texture measurements, the necessary statistical calculations were performed to determine the Pearson correlation coefficient (ρ). Correlational analysis assesses the relationship between two linear variables, with values ranging from + 1 to − 1. The sign indicates the direction of the correlation (positive or negative), while the magnitude of the coefficient reflects the strength of the relationship [14].

3 Results and Discussion

The research findings have practical implications for understanding texture behavior in relation to lateral infeed under discharge energy conditions. The study reveals that the working gap can be affected by dielectric debris and byproducts terminally promoted in different lateral infeed. Thus, harmful actions on component integrity post-electrical discharge were observed. Figure 3 illustrates the textures resulting from the WEDM process in the lateral infeed for removing Inconel 718.

To characterize the topography of Inconel 718 texture using WED machining with a discharge energy of 14.7 mJ, multifocal scanning was performed on a 5.0 × 5.0 mm area. Textures with $\Delta_y = 10$–50% are visually indistinguishable, while $\Delta_y =$ 75% presents larger crater regions. $\Delta_y = 100\%$ shows significant texture distortions. Figure 4 illustrates the textures from the WEDM process in the five lateral advances proposed for removing Stainless Steel AISI 304.

The surface texture in WEDM results from various combinations of variable parameter conditions. To better characterize the topography of the Inconel 718 and

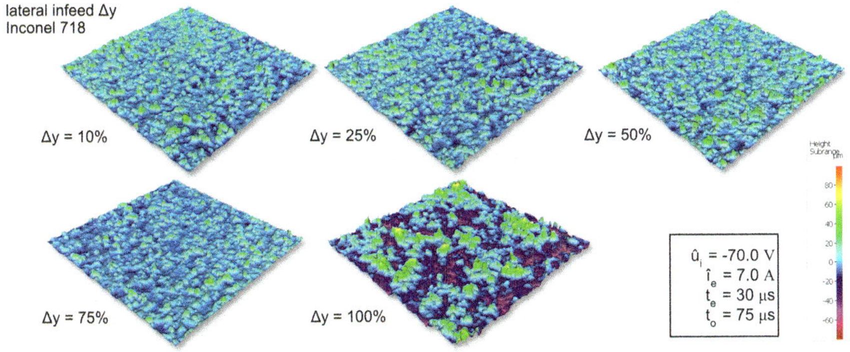

Fig. 3 Texture in the area of application of electrical discharges in Inconel 718 when changing the lateral infeed increment $\Delta_y = 10\%$ to $\Delta_y = 100\%$

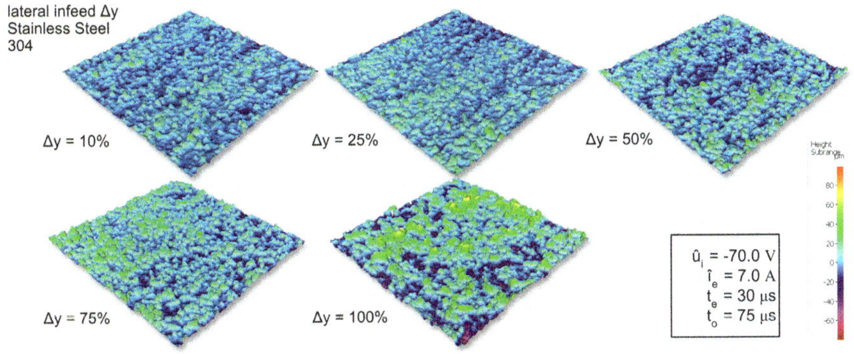

Fig. 4 Texture in the area of application of electrical discharges in Stainless Steel AISI 304 when changing the lateral infeed increment $\Delta_y = 10\%$ to $\Delta_y = 100\%$

Stainless Steel AISI 304 workpieces, 3D surface roughness parameters, specifically Sa and Sz, representing the arithmetical mean deviation of the assessed profile, are measured [4]. These parameters are widely used in industry for in-depth investigations of component texture tolerances and functional performance.

Like what was found in the topographic characterization of the texture of Inconel 718 in the WEDM removal of Stainless Steel AISI 304. The topographic quantification uses the same settings as the defined area (5.0 × 5.0 mm). It can be observed that the textures $\Delta_y = 10$–50% do not allow visually distinguishing the differences, and the condition of $\Delta_y = 100\%$ shows that the texture suffered greater distortions. To understand in numerical terms how the texture behaves, amplitude parameters Sa and Sz, as shown in Fig. 5.

The larger amplitudes shown in Fig. 5 present textures that return to the process of greater removal of the component material due to the high discharge energy. However, it is important to know the variations even in the condition of greater discharge energy

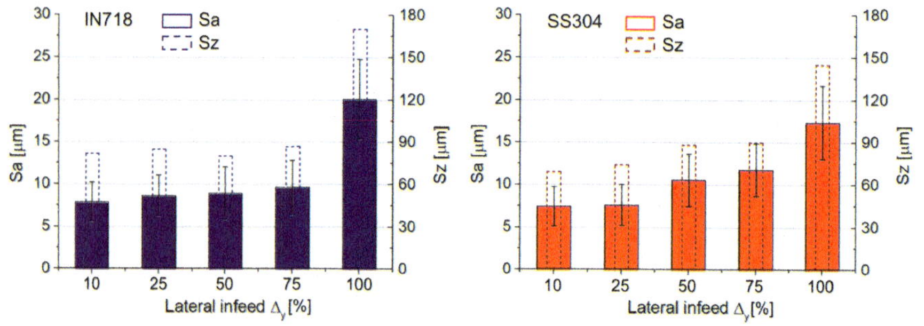

Fig. 5 Sa and Sz amplitude parameters with a lateral infeed from $\Delta_y = 10\%$ to $\Delta_y = 100\%$. On the left is the behavior of Inconel 718 (IN718), while on the right is the Stainless Steel AISI 304 (SS304)

to see the size of the damage caused by the thermal removal mechanism [15]. In Fig. 5, the roughness measurements Sa show a lower level for $\Delta_y = 10$–100%. This phenomenon occurred in both materials tested. This was possibly due to the physical properties of the sparks and the respective thermal effects on their distribution. In addition, the amount of non-uniform debris in the work flushing during WEDM machining changes as the lateral infeed rises. Thus, higher values of Sa and Sz became larger as a percentage of the wire was in action.

Kunieda and Kitamura [16] observed the distribution of discharge sites using a transparent electrode made of a single crystal of SiC. They found that the discharge sites were more widely dispersed in water when compared to hydrocarbon dielectric fluid. Additionally, they found that the radial flow of the dielectric fluid due to the discharge can propagate to greater distances from the discharge point, and the oscillation of the bubbles can continue for a longer time in water than in hydrocarbons. Thus, the study sheds light on the ignition and propagation criteria of the electrical discharge in each cycle period. However, it does not address the effects on the topography and integrity of the surface.

In studies on WEDM in Inconel 718, Carlini et al. [4] verified that the increase in the amplitude of the average surface roughness (Sa) presented greater deviations when the discharge energy was increased in both dielectrics tested. However, the lower viscosity of the deionized water resulted in better efficiency in washing the byproducts in the working gap. However, the lateral infeed was not addressed. Thus, lateral infeed studies show that the variation $\Delta_y = 10\%$ up to 75% causes little difference in roughness Sa in IN718, while for AISI 304, the variation was slightly greater but still within the error range field.

The amplitude parameter Sz followed the trend of Sa; the highest values were concentrated at $\Delta_y = 100\%$. Inconel 718 exhibited a higher Sz, approaching the wire diameter itself. This observation highlights the challenge of machining materials with high Ni–Cr content, which is less pronounced in AISI 304. The thermal conductivity properties of Inconel 718 are lower, and the efficiency of the electric discharge ends

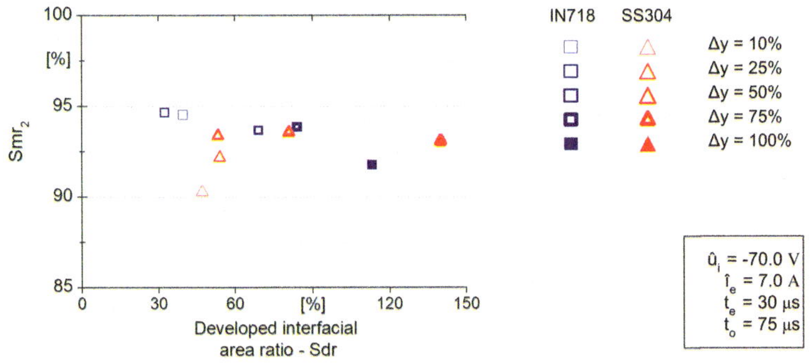

Fig. 6 Distribution and correlation parameters of Sdr and Smr_2 with lateral infeed $\Delta_y = 10\%$ to $\Delta_y = 100\%$

up being restricted. On the other hand, after the fusion is formed, the resolidification process becomes slower, and the formation of peaks and valleys is more pronounced.

In order to better understand the topography and amplitude deviations on the surface, the hybrid configurations Sdr and Smr_2 were presented and correlated. These parameters provide additional insights into the resulting topography. By quantifying the Sdr parameter in percentage terms, one can measure the increase in surface area by comparing the actual surface area with the projected area of an ideal flat surface. Therefore, higher percentage values indicate greater surface roughness. The Sdr values reflect the average magnitude of the surface irregularity gradient and increase as surface roughness intensifies [5]. Figure 6 illustrates the correlations described.

Analyzing the behavior of Sdr and Smr_2 shown in Fig. 6, it can be observed that despite the Sdr values of up to 115% for Inconel 718 and 145% for Stainless Steel AISI 304, the Areal material ratio segmenting protruding valleys from the core Smr_2. Concentrating between 90 and 95%, the variations show that the change does not influence the load-bearing region of the core in the amplitude parameters.

Lateral infeed values lower than 50% showed lower Sdr distortions. However, the variations in Smr_2 were dispersed. To compose the effect together with the two parameters, the Pearson correlation coefficient $\rho = -0.18$ was calculated, representing a weak negative correlation. Thus, the correlation index for values with large Sdr is not possible to infer or determine the correlation between the parameters. Recent studies have demonstrated that the correlation at lower discharge energy levels can lead to $\rho = -0.91$. However, in Sdr distortions of up to 15% [7].

When studying the difference in WEDM gap phenomena, the cooling efficiency of the water dielectric is significantly higher. Observation of the distribution of discharge sites shows that discharge sites are more likely to concentrate more efficiently with smaller lateral infeed. This phenomenon explains the behavior for a smaller average surface area Sa (µm).

4 Conclusions

The study investigates the surface texture of Ni–Cr–Fe-based materials manufactured using wire electrical discharge machining WEDM. Five lateral infeed rates were analyzed by varying the lateral infeed (Δ_y) relative to the wire diameter. These variations directly influenced the working face clearance, with Δ_y values set at 100, 75, 50, 25, and 10%. The topographic analysis of the effects of lateral infeed on Inconel 718 and Stainless Steel AISI 304 represents a novel contribution to the field. The conclusions drawn from the results are as follows: (i) the results show that lateral infeed affects the surface roughness Sa and Sz, respectively. Inconel 718 showed smaller variations and the largest Sz amplitude. Stainless Steel AISI 304 resulted in smaller distortions, mainly at $\Delta_y = 10\%$ and 25%; (ii) the application of $\Delta_y = 100\%$ resulted in the worst textures due to the worsening of the washing efficiency in the working gap in relation to the byproducts promoting anomalous discharge, in both materials; (iii) the Pearson correlation coefficient $\rho = -0.18$ was calculated, representing a weak negative correlation. This correlation is not characterized due to the large distortion of Sdr, and the independent variables applied in this experimental research are not present. By this analysis, operations that use the same discharge energy influence the lateral infeed are reduced, mainly in the conditions of full removal.

References

1. Reolon LW, Laurindo CAH, Torres RD, Amorim FL (2019) WEDM performance and surface integrity of Inconel alloy IN718 with coated and uncoated wires. Int J Adv Manuf Technol 100:1981–1991
2. Seshaiah S, Sampathkumar D, Mariappan M, Mohankumar A, Balachandran G, Kaliyamoorthy M, Rajendran B, Gopal R (2022) Optimization on material removal rate and surface roughness of stainless steel 304 wire cut EDM by response surface methodology. Adv Mater Sci Eng 2022
3. Suárez A, Veiga F, Polvorosa R, Artaza T, Holmberg J, de Lacalle LNL, Wretland A (2019) Surface integrity and fatigue of non-conventional machined alloy 718. J Manuf Process 48:44–50
4. Carlini GC, da Silva C, Torres RD, Soares P, Weingaertner WL, Amorim FL (2022) WED-machining performance by reciprocating molybdenum wire on Inconel 718 with water or hydrocarbon dielectrics. Int J Adv Manuf Technol 119:1853–1866
5. Pagani L, Qi Q, Jiang X, Scott PJ (2017) Towards a new definition of areal surface texture parameters on freeform surface. Meas J Int Meas Confed 109:281–291
6. Qi Q, Li T, Scott PJ, Jiang X (2015) A correlational study of areal surface texture parameters on some typical machined surfaces. Procedia CIRP 27:149–154
7. Carlini GC, dos Santos Roik I, Blödorn R, Torres RD, Amorim FL (2024) Beyond the spark: the impact of tool electrode materials, polarities, and pulse distribution on machined surface texture in the electrical discharge machining of Inconel 718 superalloy. Int J Adv Manuf Technol 133:2951–2965
8. Buk J (2022) Surface topography of Inconel 718 alloy in finishing WEDM. Adv Sci Technol Res J 16:47–61

9. VDI (1990) VDI 3402. Elektroerosive bearbeitung: definitionen und terminologie. Deutsches Institut für Normung, Düsseldorf
10. Ezugwu EO, Wang ZM, Machado AR (1998) The machinability of nickel-based alloys: a review. J Mater Process Technol 86:1–16
11. Naeim N, AbouEleaz MA, Elkaseer A (2023) Experimental investigation of surface roughness and material removal rate in wire EDM of stainless steel 304. Materials 16(3). https://doi.org/10.3390/ma16031022
12. Johny A, Thiagarajan C (2020) Investigation of surface integrity and it is optimization on pure titanium using molybdenum wire by reciprocated travelling WEDM—a review. Mater Today Proc 33:2581–2584
13. Czifra Á, Barányi I (2020) SDQ-SDR topological map of surface topographies. Front Mech Eng 6:1–9
14. Guo P, Xiong Z, Dong Z, Zhang S, Piano S, Liu M (2024) Multi-parameterised surface texture characterisation for ultra-precision machined surfaces. Surf Topogr Metrol Prop 1–23
15. Klocke F, König W (2007) Fertigungsverfahren 3. Abtragen, Generieren und Lasermaterialbearbeitung. Springer Berlin Heidelberg, Berlin
16. Kunieda M, Kitamura T (2018) Observation of difference of EDM gap phenomena in water and oil using transparent electrode. Procedia CIRP 68:342–346

Energy Efficiency in Machine Tools: Impact of Cutting Parameters on Sinusoidal Paths

Marie Madeleine Sarzi Inacio, Rodrigo Henriques Lopes da Silva, and Amauri Hassui

Abstract This study aimed to investigate the impact of cutting parameters, cutting speed (v_c), depth of cut (a_e), and feed per tooth (f_z), on energy consumption, active power, and energy efficiency during the milling of Ti-6Al-4V alloy with sinusoidal tool paths on CNC machine tools. The methodology included experimental trials with energy data collection at the CNC machine input using a power quality analyzer. A 23 factorial experimental design was employed to define machining parameters for the tests. Experimental results showed that increasing in the cutting speed and depth of cut significantly raised the mean active power and energy consumption. However, operating at higher cutting speeds and feed per tooth resulted in optimized energy efficiency, with the most efficient configuration being v_c of 85 m/min, f_z of 0.16 mm/tooth, and a_e of 1.0 mm, achieving approximately 27.5% greater efficiency compared to the least efficient configuration. These findings highlight the importance of selecting appropriate cutting parameters to maximize energy efficiency in machining processes, contributing to sustainable manufacturing practices.

Keywords Energy efficiency · CNC machining · Cutting parameters · Sinusoidal path · Ti-6Al-4V

1 Introduction

The manufacturing industry plays a significant role in global development, accounting for a substantial portion of industrial energy consumption in many countries and contributing significantly to carbon dioxide (CO_2) emissions from production processes [1]. Given the challenges of climate change and environmental degradation, re-evaluating production processes has become essential. Bunse et al. [2]

M. M. S. Inacio (✉) · R. H. L. da Silva
Federal University of Technology—Paraná, Cornélio Procópio, Brazil
e-mail: marie.sarzi@hotmail.com

A. Hassui
Faculty of Mechanical Engineering, University of Campinas, Campinas, Brazil

M. C. Gomes et al. (eds.), *Innovations in Brazilian Machining*, Lecture Notes in Mechanical Engineering, https://doi.org/10.1007/978-3-031-92819-2_21

emphasize that, with increasing ecological awareness, industries face substantial challenges.

Sustainable manufacturing, which balances environmental impact and economic feasibility, has become a strategic priority for companies focused on innovation and regulatory compliance. Moreover, global attention and new guidelines are increasingly being directed toward this topic. Sustainable practices are essential to minimize environmental impact and promote long-term ecological health [3–5]. Consequently, addressing resource consumption, including electrical energy, has become a critical priority in adopting sustainable practices.

For instance, global industrial energy demand accounted for approximately 38% of total energy generated in 2018 [6], underscoring the urgency of developing solutions to minimize environmental impacts and promote sustainability through advanced technologies and enhanced process analyses. Industrial energy consumption has steadily increased at an annual rate of about 1.3% between 2010 and 2022 [7]. International Energy Agency [8] highlights the importance of improving energy efficiency as costs rise and supply disruptions threaten industrial operations. Consequently, urgent measures and policies have been implemented by both consumers and governments to conserve and manage energy consumption more effectively.

Electrical energy consumption in machine tools is a key concern, particularly in the metalworking industry. This sector accounts for 1–3% of global electricity demand [9], with machine tools playing a significant role in this consumption. Thus, finding solutions to reduce energy use has become increasingly relevant. According to Soori et al. [10], optimizing machining processes is essential to achieving energy efficiency, emphasizing the importance of machining parameters and operations.

Machine tools are widely recognized for their substantial energy losses and CO_2 emissions. Despite the associated implementation costs, this characteristic and their significant energy demands make them a logical starting point for energy efficiency initiatives in the manufacturing industry [11].

Recent studies have explored these challenges. For example, Chen et al. [12] examined how optimizing parameters and tool paths can minimize material waste and energy consumption, reduce machining time, and extend tool life [10] and noted that optimizing tool paths and selecting suitable tools can also reduce coolant usage.

Other works Jamil et al. [11] and Khan et al. [13] have focused on energy characterization and strategies for energy savings in sustainable machining processes, Jamil et al. [11] specifically examined the trade-offs between energy efficiency, carbon emissions, and machining characteristics in milling Ti-6Al-4V.

In this context, optimizing energy consumption in machining operations, particularly in CNC machine tools, presents a significant opportunity to improve energy efficiency and reduce environmental footprints.

Thus, this study investigates the impact of cutting parameters, specifically cutting speed (v_c), depth of cut (a_e), and feed per tooth (f_z), on electrical active power, energy consumption, and energy efficiency during machining with sinusoidal tool paths on CNC machine tools. This analysis aims to identify how these variables influence process energy efficiency, seeking to optimize operations and contribute to more sustainable manufacturing practices.

2 Methodology

This study employed a structured methodological approach, encompassing the planning and execution of experimental trials. The testing methods, machining parameters, and boundary conditions were analyzed and documented to generate relevant results. The trials, performed on a CNC machine using the Ti-6Al-4V titanium alloy, focused on evaluating the impact of cutting parameters (v_c, a_e, f_z) on sinusoidal tool paths, with electrical energy data collected for subsequent analysis.

2.1 Machine Tool, Material, Cutting Tool, and Data Acquisition Equipment

Milling trials of the Ti-6Al-4V titanium alloy were conducted on a vertical machining center, model DISCOVERY 4022, manufactured by ROMI. This equipment is located in SENAI's Industry 4.0 Plant, dedicated to Research and Development projects at the Metalworking Institute in São Paulo state, Brazil.

The machine's maximum spindle speed is 6000 RPM, featuring an ISO 40 taper tool holder and a MAC 9 numerical control system.

The test specimens were provided by a research group conducting previous studies on the effects of distinct cutting strategies on surface finish, cutting forces, and tool life [14, 15]. The fixture device, designed and manufactured by the same research group, was made available to clamp the test specimens securely during the experimental trials. The fixture measures 170 × 100 × 53 mm and includes four 8.5 mm diameter holes on the main base. Its surface has 24 M6 threaded holes, enabling the specimens to be clamped in various positions during the tests. The test specimens measured 220 × 85 × 10 mm and were perforated with four 6.5 mm through-holes to accommodate four M6 screws.

Solid carbide end mills from Sandvik Coromant, classified as grade S (R216.25-25050OCL54P1620), were used. These tools have a 25 mm diameter, a maximum cutting-edge length of 54 mm, and five cutting edges. Additionally, they are coated with TiAlN, applied via Physical Vapor Deposition (PVD), which enhances wear resistance at high temperatures. A SECO hydraulic chuck, model BT40, HC 25 × 105 mm, was used to clamp the end mills.

A cutting fluid, HOCUT 795 H, manufactured by Quaker Houghton, was used to mitigate the heat effects on the tool. This semi-synthetic, bio-resistant fluid is suitable for machining ferrous and non-ferrous metals. A 1:40 dilution and 5% refractometer reading were used, with the fluid applied externally at a flow rate of approximately 16 L/min.

The Fluke 435 Series II—POWER QUALITY AND ENERGY ANALYZER was used to acquire energy parameters from the CNC machining center (general energy input of the machine tool). This robust instrument is designed for analyzing and diagnosing power quality in three-phase systems.

2.2 Cutting Parameters and Experimental Design

Following an extensive literature review, the influence of certain variables on tangential milling of the Ti-6Al-4V alloy with a sinusoidal path was investigated. The selected variables included cutting speed (v_c), feed per tooth (f_z), and depth of cut (a_e). These variables were chosen based on previous studies [14, 15], which highlighted their importance in machining processes, particularly regarding surface finish, cutting forces, and tool life, core elements of this research.

Based on these variables and methodology, a workflow was developed to conduct the experiments. A 23 factorial design was applied to experiments involving the sinusoidal path, considering v_c (60; 85 m/min), f_z (0.11; 0.16 mm/tooth), a_e (0.5; 1 mm). Subsequently, the mean active power (W) was measured. Following this, the maximum machining parameters ($v_c = 85$, $f_z = 0.16$, $a_e = 1$) were applied to measure the instantaneous active power (W), energy consumption (Wh), and energy efficiency (mm^3/Wh), enabling the analysis of results obtained from the sinusoidal milling path.

A factorial experimental design method with multiple factors was chosen, maintaining a controlled number of repetitions. This decision was driven by the high cost of the Ti-6Al-4V alloy used in the project, execution time, and equipment availability. The factorial method is also robust and reliable, optimizing resource and time usage.

For all experiments, down milling was employed. Continuous cooling was used during operations, and the sinusoidal path was defined with a period of 60 mm and an amplitude of 30 mm, ensuring the tool's cutting edge remained within its usable range of 54 mm.

A dedicated software was used to analyze the collected data for statistical analysis of the experimental results. Energy data, including active power and energy consumption, were recorded from the CNC machine's main power input, providing a foundation for analyzing and discussing the results.

2.3 Output Variables

The output variables analyzed in this study include active power, energy consumption, and energy efficiency. All were obtained and calculated from data collected using the Fluke 435 Series II—POWER QUALITY AND ENERGY ANALYZER.

Active power (P) represents the energy effectively used to work on the CNC machine during machining. This variable was obtained from the energy analyzer data and subsequently used in calculations. Furthermore, it can be calculated as described by Alexander and Sadiku [16] using Eq. (1), where V is the voltage, I the current, and $\cos(\phi)$ the power factor, all measured by the energy analyzer.

$$P = V \times I \times \cos(\phi) \tag{1}$$

Energy consumption (E) was calculated by multiplying total power P_T by the operation time t, according to Eq. (2), allowing the analysis of the total energy consumed during the process [16].

$$E = P_T \times t = Wh \tag{2}$$

Energy efficiency is defined as the ratio of useful energy (production) to total energy consumption, as discussed by Refs. [17, 18]. This variable reflects the effectiveness of energy use during machining, focusing on material removal over the machining period (mm³/Wh). Energy efficiency was calculated using Eq. (3).

$$EE = Useful\ production/Total\ energy\ consumption = \mathrm{mm}^3\ /\mathrm{Wh} \tag{3}$$

3 Results and Discussions

The procedures carried out in this study were divided into phases, and this section presents and discusses the results obtained for each of the established phases.

For all analysis phases, the data exhibited normal distribution patterns, and measurement variations remained within expected limits, which allowed for the application of analysis of variance (ANOVA) to statistically process the results observed in the study.

3.1 Active Power Analysis: Sinusoidal Milling Path

The effects of interactions and correlations between machining parameters and the mean active power were evaluated in an initial series of experiments (23 factorial design). Data were collected at one-second intervals during the sinusoidal milling of the Ti-6Al-4V alloy. The three factors analyzed included cutting speed (v_c: 60 and 85 m/min), depth of cut (a_e: 0.5 and 1.0 mm), and feed per tooth (f_z: 0.11 and 0.16 mm/tooth).

The main effects of the parameters on mean active power were analyzed. It was observed that the mean active power (P) increased from approximately 1955 W to about 1990 W when the cutting speed was raised from 60 to 85 m/min. This indicates a direct relationship between mean active power and higher v_c during sinusoidal milling, further corroborated by a statistically significant p-value.

A direct correlation was also noted with depth of cut, where mean active power increased from approximately 1952–1994 W as a_e rose from 0.5 to 1.0 mm. The a_e increase was the most significant among the three factors, indicating that a_e had the greatest impact on mean active power in this scenario, with a p-value of 0.015 confirming its significance.

The increase in mean active power varying feed per tooth was minimal compared to the other two factors. Additionally, feed per tooth (f_z) was not significant for mean active power variation in sinusoidal milling, as indicated by its p-value.

Subsequently, instantaneous active power data from the CNC machine was analyzed over time for the sinusoidal milling process, with data collected every second. The extreme parameters were applied (v_c of 85 m/min, f_z of 0.16 mm/tooth, and a_e of 1 mm). The results illustrate the typical behavior of instantaneous active power observed throughout the study.

In the first 8 s, instantaneous active power increased rapidly, reaching approximately 2150 W, reflecting the initial moment when the tool contacts the material and begins cutting. During the cutting process, a stabilization of around 2200 W between 8 and 29 s was observed, and minor variations in power were evident. Generally, instantaneous active power fluctuated between 2100 and 2250 W during sinusoidal milling, likely due to cutting characteristics and the operation of CNC machine components using the sinusoidal path.

Toward the end, a peak in instantaneous active power was noted, reaching approximately 2650 W, followed by a sharp drop at 36 s, indicating the end of the process as the tool disengaged from the Ti-6Al-4V alloy. The variations in instantaneous active power along the sinusoidal path illustrate a noticeable sinusoidal trend in the collected power data during material-tool contact.

Regarding energy consumption, it was observed that increasing the cutting speed from 60 to 85 m/min reduced energy consumption, indicating that operating at higher cutting speeds for this process condition results in more efficient energy consumption. Additionally, cutting speed and feed per tooth were significant parameters for energy consumption, as evidenced by their p-values.

The analysis reveals that increasing cutting speed clearly enhances process efficiency by reducing energy consumption. Similarly, increasing feed per tooth also improved energy consumption. Although the depth of cut had a lesser impact and was not a significant parameter, increasing ae from 0.5 to 1 mm resulted in a slight increase in energy consumption.

3.2 Energy Efficiency Analysis

The energy efficiency results revealed that a cutting speed of 60 m/min and a feed per tooth of 0.11 mm/tooth is not the most energy-efficient combination for the process. However, energy efficiency improves when feed per tooth is increased to 0.16 mm/tooth while maintaining a cutting speed of 60 m/min. Thus, combined with a depth of cut (a_e) of 1 mm, this configuration achieved approximately 92 mm^3/Wh.

Moreover, the results showed that combining a higher cutting speed (85 m/min) with a smaller feed per tooth (0.11 mm/tooth) and a greater depth of cut (1.0 mm) resulted in elevated energy efficiency. However, the configuration that achieved the maximum energy efficiency in the CNC machine combined a high cutting speed

(85 m/min), a higher feed per tooth (0.16 mm/tooth), and a greater depth of cut (1 mm), as presented in Fig. 1.

The findings emphasize that increasing cutting speed improves the CNC machine's energy efficiency, especially when combined with a higher feed per tooth. Another key factor is that a higher feed per tooth enhances the machine's energy efficiency, regardless of the depth of cut within the tested range of cutting conditions. Furthermore, a depth of cut of 1 mm, combined with the optimized parameters, clearly results in a process with improved energy efficiency during machine operation compared to other tested parameter combinations.

Finally, the configuration with v_c of 85 m/min, f_z of 0.16 mm/tooth, and a_e of 1.0 mm achieved the highest energy efficiency (27.5% greater than the least efficient condition), making it the ideal configuration for maximizing the CNC machine's energy efficiency in terms of energy consumption.

Research by Inácio et al. [14] demonstrates that combining an appropriate tool path, such as a sinusoidal one, with the correct cutting parameters can significantly improve surface finish (by over 130%), reduce cutting forces (by approximately 20%), and alter wear mechanisms, extending tool life by 4–5 times without compromising productivity. The parameters used in this study align with the methodology of [14], which is particularly relevant within a sinusoidal path. Finally, the findings highlight the potential to select machining parameters that increase CNC machine energy efficiency and improve other aspects of the machining process.

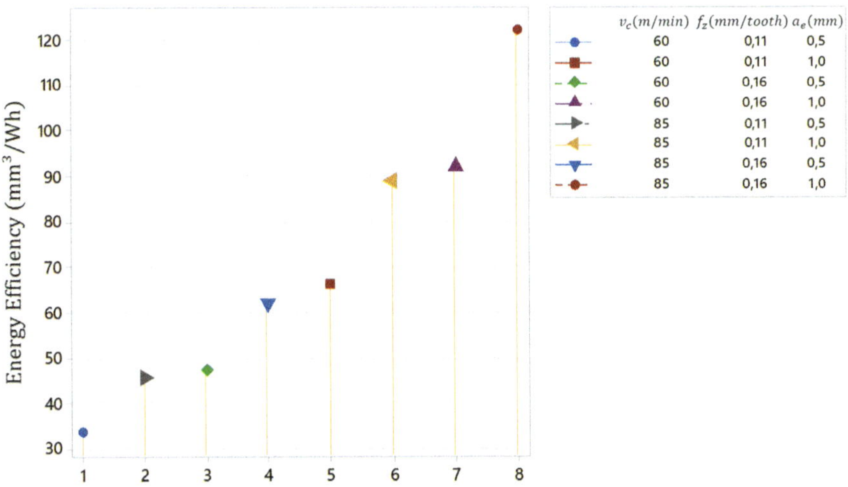

Fig. 1 Energy efficiency achieved in experiments with the applied parameters

4 Conclusions

The results of this study show that increasing cutting speed from 60 to 85 m/min raises the mean active power, indicating a direct and significant relationship with energy consumption. Depth of cut (a_e) was found to have the greatest influence, while feed per tooth (f_z) had minimal effect.

The analysis of instantaneous active power over time revealed typical variations during sinusoidal milling, with initial and final peaks as well as stabilization during the cutting phase. Regarding energy consumption, operating with higher cutting speeds and feed per tooth resulted in greater CNC machine energy efficiency.

The standout configuration for energy efficiency, approximately 27.5% higher than the least efficient condition found in the experiments, was achieved with a cutting speed of 85 m/min, a feed per tooth of 0.16 mm/tooth, and a depth of cut of 1.0 mm. This combination was identified as the most efficient for optimizing the equipment's energy consumption during sinusoidal milling, striking an ideal balance between productivity and energy usage. This result is attributed to the direct impact of operation time and mean active power on electrical energy consumption, consequently affecting the process's energy efficiency.

Furthermore, as validated in previous studies, the process parameters applied in this study extend tool life by 4–5 times without affecting productivity.

These results, aligned with existing studies, underscore the feasibility of optimizing energy consumption in CNC machines without compromising machining quality, highlighting the importance of precise adjustments to cutting parameters.

References

1. Gunjal S, Sanap S (2020) A review on some investigations into sustainable machining techniques. E3S Web Conf 170:2003
2. Bunse K, Vodicka M, Schönsleben P, Brülhart M, Ernst FO (2011) Integrating energy efficiency performance in production management—gap analysis between industrial needs and scientific literature. J Clean Prod 19(6):667–679
3. Alves W, Silva Â, Rodrigues HS (2020) Green practices as a path towards sustainability: evidence from Portuguese companies. Bus Syst Res J 11(2):7–20
4. Duarte S, Cabrita MDR, Cruz-Machado V (2023) Green supply chain practices: toward a sustainable industry development. In: Xu J, Windapo A, Hassan MHA, Hajiyev A (eds) E3S web of conferences, vol 409. EDP Sciences, Les Ulis, p 1004
5. Elleuch B, Bouhamed F, Elloussaief M, Jaghbir M (2018) Environmental sustainability and pollution prevention. Environ Sci Pollut Res 25(19):18223–18225
6. International Energy Agency (2019) Energy efficiency 2019. IEA, Paris
7. International Energy Agency (2023) Energy efficiency 2023. IEA, Paris
8. International Energy Agency (2022) Energy efficiency 2022. IEA, Paris
9. Denkena B, Abele E, Brecher C, Dittrich M, Kara S, Mori M (2020) Energy-efficient machine tools. CIRP Ann 69(2):646–667
10. Soori M, Jough FKG, Dastres R, Arezoo B (2024) Sustainable CNC machining operations: a review. Sustain Oper Comput 5:73–87

11. Jamil M, Zhao W, He N, Gupta MK, Sarikaya M, Khan AM, Sanjay MR, Siengchin S, Pimenov DY (2021) Sustainable milling of Ti–6Al–4V: a trade-off between energy efficiency, carbon emissions and machining characteristics under MQL and cryogenic environment. J Clean Prod 281:125374
12. Chen R, Shen H, Lai Y (2022) A metaheuristic optimization algorithm for energy efficiency in digital twins. Internet Things Cyber-Phys Syst 2:159–169
13. Khan AM, He N, Jamil M, Raza SM (2021) Energy characterization and energy-saving strategies in sustainable machining processes: a state-of-the-art review. J Prod Syst Manuf Sci 2(1):3–3
14. Inácio R, Da Silva RL, Pereira I, Hassui A (2022) Suppressing notch wear by changing the tool path in the side milling of a Ti-6Al-4V alloy. Int J Adv Manuf Technol 125:453–463
15. Da Silva RHL, Hassui A (2018) Cutting force and surface roughness depend on the tool path used in side milling: an experimental investigation. Int J Adv Manuf Technol 96(1–4):1445–1455
16. Alexander CK, Sadiku MNO (2013) Fundamentals of electric circuits with applications, 5th edn. AMGH Editora, Porto Alegre
17. International Organization for Standardization: ISO 14955-1:2017 (2017) Machine tools—environmental evaluation of machine tools—Part 1: design methodology for energy-efficient machine tools. ISO, Geneva. https://www.iso.org/standard/70035.html. Last accessed 25 Jun 2024
18. Tanaka K (2008) Assessment of energy efficiency performance measures in industry and their application for policy. Energy Policy 36(8):2887–2902

The manufacturer's authorised representative in the EU is Springer Nature Customer Service Centre GmbH, Europaplatz 3, 69115 Heidelberg, Germany. If you have any concerns regarding our products, please contact ProductSafety@springernature.com

Printed and bound by CPI Group (UK) Ltd, Croydon, CR0 4YY
30/09/2026
02225604-0019